JSC-66136
Revision 1

Space Shuttle Guidance, Navigation, and Rendezvous Knowledge Capture Reports

Mission Operations Directorate
Flight Dynamics Division

October 2011

This document has been reviewed for Proprietary, SBU, and Export Control (ITAR/EAR) and has been determined to be nonsensitive. It has been released to the public via the NASA Scientific and Technical Information (STI) Process DAA 24818.

National Aeronautics and
Space Administration

Lyndon B. Johnson Space Center
Houston, Texas

This page intentionally left blank.

JSC-66136
Revision 1

Space Shuttle Guidance, Navigation, and Rendezvous Knowledge Capture Reports

October 2011

By:

Original Signed by:

John L. Goodman

*Flight Design and Dynamics
United Space Alliance, LLC*

Approved By:

Original Signed By:

Christine M. Reichert

*Chief
Orbit Dynamics Branch
Flight Dynamics Division
Mission Operations Directorate
NASA Johnson Space Center*

This page intentionally left blank.

Table of Contents

Preface 7

Introduction 8

Observations on Documentation Quality and Research Challenges 11

    Programmatic Histories Compared to Technical Documentation 12

    Relative Quality of Apollo and Space Shuttle Documentation 14

    Apollo and Space Shuttle Research Challenges 17

    Knowledge Capture via Software Documentation 21

    Orion Knowledge Capture Challenges 26

Space Shuttle Knowledge Capture Approach 30

    The Knowledge Capture and Management Process 31

    Leading and Managing People is the Key to Effective Knowledge Capture 33

    Documenting Lessons Learned and Experiences 35

    Choosing Documentation to Preserve 38

Lessons Learned and Best Practices Reports 41

    Lessons Learned From Seven Space Shuttle Missions 42

    Flight Safety Education for Current and Future Space Flight Programs 43

    Best Practices for Researching and Documenting Lessons Learned 44

    Knowledge Capture and Management for Space Flight Systems 45

Navigation and GPS Reports 46

    Space Shuttle GPS Operating Characteristics 47

    Space Shuttle GPS Compilation 48

    Improvement of Space Shuttle Time to Node Computation 49

    Space Shuttle Rotation, Nutation and Procession (RNP) Matrix Computation 50

    GPS Lessons Learned From the ISS, Space Shuttle, and X-38 51

    Navigation Technical History with Lessons Learned 55

    Operational Aspects of Space Shuttle GPS 57

    Ionopheric Scintillation and GPS 58

    Navigation Papers by Gene Brownd and Dr. Leonard Kramer 59

Rendezvous and Proximity Operations Reports 60

    Space Shuttle Rendezvous and Proximity Operations Experience Report 61

    Rendezvous and Proximity Operations Systems Integration 64

    Space Shuttle Rendezvous Training 65

Approved for public release via STI DAA 24818. See statement on title page.

Visiting Vehicles Rendezvous Training 67
   Compilation Volumes of Rendezvous and Proximity Operations Papers 69
   History of Space Shuttle Rendezvous 71
Guidance and Targeting Reports 73
   Lambert Targeting Compilation 74
   Lambert Cyclic Guidance Compilation 74
   STS-49 Lambert Targeting Compilation 74
   Lambert Cyclic Guidance Upgrade 75
   Compilation of Powered Explicit Guidance Papers 76
Apollo and Orion Reports 78
   Apollo 13 Guidance, Navigation, and Control Challenges 79
   Introduction to Apollo Entry Guidance and Flight Performance 80
   Challenges of Orion Rendezvous Development 81
   Orion On-Board Navigation Architecture and Operations Concepts 82
Compilations of Draper Laboratory Papers 83
Summary 84

Preface

This document is a catalog and readers guide to lessons learned, experience, and technical history reports, as well as compilation volumes prepared by United Space Alliance personnel for the NASA/Johnson Space Center (JSC) Flight Dynamics Division.[1] It is intended to make it easier for future generations of engineers to locate knowledge capture documentation from the Shuttle Program.

The first chapter covers observations on documentation quality and research challenges encountered during the Space Shuttle and Orion programs. The second chapter covers the knowledge capture approach used to create many of the reports covered in this document. These chapters are intended to provide future flight programs with insight that could be used to formulate knowledge capture and management strategies.

The following chapters contain descriptions of each knowledge capture report. The majority of the reports concern the Space Shuttle. Three are included that were written in support of the Orion Program. Most of the reports were written from the years 2001 to 2011. Lessons learned reports concern primarily the shuttle Global Positioning System (GPS) upgrade and the knowledge capture process. Experience reports on navigation and rendezvous provide examples of how challenges were overcome and how best practices were identified and applied. Some reports are of a more technical history nature covering navigation and rendezvous. They provide an overview of mission activities and the evolution of operations concepts and trajectory design. The lessons learned, experience, and history reports would be considered secondary sources by historians and archivists.

Compilation volumes preserve key historical documents (memos, presentations, reports) covering powered flight guidance, rendezvous, and navigation. These would be considered primary sources by historians and archivists. Document overviews in each volume provide context for understanding the primary source materials.

The reports presented in this document are NASA contractor reports, JSC documents, publically available conference papers and journal articles. NASA contractor reports may be obtained from the NASA Technical Reports Server (NTRS) on the internet. JSC documents are available from the JSC Scientific Technical and Information Center (STIC, mail code IS23) in Building 45 and may be ordered through the technical library at any NASA center.[2] Conference papers and journal articles can be obtained from the sponsoring professional organization (AAS, AIAA, IEEE, ION) websites or through libraries at NASA centers and universities.[3]

---

[1] At the time of publication the Mission Operations Directorate (MOD) Flight Dynamics Division (code DM) performed mission planning, trajectory design, some crew training, and real-time mission support for the Space Shuttle and International Space Station (ISS) Programs. It also performed mission planning and trajectory oversight, as well as real-time mission support, for ISS Visiting Vehicles (Soyuz, Progress, ATV, HTV, Dragon, Cygnus, etc.).

[2] Lyndon B. Johnson Space Center, National Aeronautics and Space Administration, 2101 NASA Parkway, Houston, TX 77058-3696

[3] AAS – American Astronautical Society, AIAA – American Institute of Aeronautics and Astronautics, IEEE – Institute of Electrical and Electronics Engineers, ION – Institute of Navigation.

Approved for public release via STI DAA 24818. See statement on title page.

Introduction

The near simultaneous terminations of the Space Shuttle and Constellation Programs presents a significant challenge for ensuring the preservation and continuity of NASA's human space flight brain trust. "How do we preserve and rebuild the brain trust for future flight programs?" is a question that many within NASA have been asking. Future programs will not exactly duplicate the technical and flight technique approaches of the Space Shuttle, but much of the knowledge and experience of shuttle personnel will be valuable for new programs to reduce technical, cost, and schedule risk.

The 2010 report of the Aerospace Safety Advisory Panel (ASAP) expressed this concern,

> *"NASA has deep knowledge built over many years on how to put humans into space, and this knowledge could be very useful to commercial developers. There is excellent work being done across the spectrum of the Shuttle, the ISS, and Constellation. It will be a great loss to the Nation and to humankind if this knowledge is not captured, managed, and effectively utilized."*[1]

This is not a new concern. In the introduction to the 1967 NASA publication *Scientific Satellites* Deputy Associate Administrator (Sciences) John E. Naugle wrote,

> *"There is grave danger that the line of development of space equipment and instrumentation may be lost if care is not given to its preservation. Much information is contained in in-house reports, but, as in all active fields, the records are scattered, often incomplete, and sometimes silent on important points. In time, personnel, too, can be expected to begin to scatter. While it can still be recovered, it is important that this information be recorded. Otherwise, in the future much of the usefulness of present-day measurements could be jeopardized. As future investigators try to assess past results and to combine them with their own, they will need to know accurately how the results were obtained."*[2]

Effective knowledge capture and management to address corporate knowledge loss concerns is beneficial for engineers and managers working on the design, development, test, and engineering (DDT&E) phase of a program as well as during the planning and execution of missions. In addition to mitigating the risk of corporate knowledge loss, effective knowledge capture and management can also enable engineers and management to identify trends of potentially anomalous spacecraft systems performance that could result in the failure to meet mission objectives or loss of the vehicle and crew. Loss of both *Challenger* and *Columbia* were due to a lack of understanding of hardware performance. In the case of *Challenger* it was Solid Rocket Booster o-ring performance at cold temperatures, and in the case of *Columbia* it was External Tank foam shedding.

---

[1] *Aerospace Safety Advisory Panel Annual Report for 2010*, NASA, Washington, DC, January 2011.
[2] Corliss, William R., *Scientific Satellites*, NASA-SP-133, NASA, Washington, DC, 1967.

Approved for public release via STI DAA 24818. See statement on title page.

Introduction

Allan J. McDonald, the senior Morton Thiokol representative at the STS-51L Flight Readiness Review (FRR) held at the Kennedy Space Center recommend that *Challenger* not be launched due to cold weather. He later wrote:

> *"I wasn't recommending not launching because of what I knew, but because of what I didn't know, and I thought that NASA was in the same position. It just wasn't worth taking the risk with all these unknowns."*[3]

Technical discoveries are more likely to be made or recognized by intellectually curious personnel that seek understanding of systems performance and theory underlining system and software design. These discoveries are typically made when spacecraft systems are operated at the extremities of the performance envelop, when understanding of systems performance is recognized as being inadequate, or when interactions between hardware and software elements of spacecraft systems are not fully understood. Technical discoveries are learning experiences that should be followed by explicit and tacit knowledge capture so that knowledge gained is preserved and passed on to current and future personnel.[4]

During the DDT&E phase of a flight program engineers may evaluate legacy flight programs for algorithms, hardware designs, integration architectures, and mission design techniques that could be re-used to reduce cost, schedule, and system development risk. Well-written documentation from legacy programs enables spacecraft DDT&E personnel to determine if legacy technical solutions are applicable to a future flight program. Furthermore, system design rationale and history from legacy programs may enable DDT&E personnel to anticipate technical and performance challenges, and write more robust and realistic requirements.

Effective knowledge capture and management during the DDT&E phase can reduce risk and increase the likelihood of mission success during the flight phase. Ensuring safety and mission success depends on development, verification, performance analysis, and maintenance of hardware and software in on-board systems, ground systems, and ground facilities. Extensive analysis is performed in support of mission design, procedure development, and hardware evaluation. These activities require insight into underlying theory, requirements rationale, analysis techniques, systems performance and modification history, and software tools over the life of a program. The increasing complexity and proliferation of computer networks in on-board and ground systems necessitates insight into software design and operation.[5]

Many engineers working on the operations side of a flight program may not have participated in the development and certification of the spacecraft hardware and software they are concerned with. As a result, they may lack insight into systems design rationale. Some insight may be obtained through social networking with more knowledgeable engineers or from preserved

---

[3] McDonald, Allan J., with James R. Hansen, *Truth, Lies, and O-Rings*, University Press of Florida, Gainesville, FL, 2009, page 4.
[4] *Proceedings of the F-8 Digital Fly-By-Wire and Supercritical Wing First Flight's 20th Anniversary Celebration*, May 27, 1992, NASA Dryden Flight Research Center, NASA Conference Publication 3256, Volume I, page 16.
[5] Goodman, John L., *Knowledge Capture and Management for Space Flight Systems*, NASA Contractor Report NASA/CR-2005-213692, NASA Johnson Space Center, October 2005.
Approved for public release via STI DAA 24818. See statement on title page.

documentation (if documentation was prepared in the first place). However, over time knowledge loss occurs as key personnel leave flight programs, corporate and government agency reorganizations occur, or as documentation is lost or purposely discarded. The end of flight programs and loss of personnel through layoffs can destroy a brain trust that has been maintained and nurtured for years or decades.

Aerospace text books and technical papers in the open literature provide limited coverage of algorithms, hardware, and mission design techniques as actually applied to launch vehicles and spacecraft. Even if an algorithm or hardware unit has demonstrated excellent performance over the life of a flight program, a thorough understanding of it is still required to address performance anomalies or apply the algorithm or hardware to new flight programs.

Corporate knowledge loss negatively impacts the ability of engineers to perform accurate analyses in a timely manner. Significant amounts of time may be expended in an attempt to understand analyses performed and technical decisions made in the past. In some cases, lack of insight may force an analysis to be completely redone. Incomplete understanding of system requirements rationale, underlying design theory, and systems performance history degrades the quality of engineering work. Corporate knowledge loss also makes it difficult for engineers to understand, evaluate, modify and reuse software years or decades after it was written and certified. The same is true for hardware and ground facilities. The result is increased life cycle costs and risk to safety and mission success.

Effective mentoring and access to key historical documentation for second, third, fourth, and subsequent generations of engineers is critical in an industry with a turnover rate and little margin for error. Creation and preservation of informative and well-written documentation is a necessary for ensuring good contract performance and mission success.[5]

In his memoir *Truth, Lies, and O-Rings* former Morton Thiokol engineering manager Allan J. McDonald identified the primary lesson from the *Challenger* and *Columbia* accidents by writing:

> *"I sincerely hope that we will have learned the powerful lesson that the Challenger and Columbia accidents should have taught us all. That lesson is that problems in technically complex systems must be addressed with total honesty by highly competent scientists and engineers of high integrity. There is no place for bureaucratic solutions to technical problems, and any attempt to do so should be resisted by ethical engineers and scientists."*[6]

Addressing technical issues in this manner requires curious personnel who rigorously pursue better understanding of vehicle systems performance and identification of potential risks. Effective knowledge capture and management is necessary to enable and support such investigations.

---

[6] McDonald, Allan J., with James R. Hansen, *Truth, Lies, and O-Rings*, University Press of Florida, Gainesville, FL, 2009, page 563.
Approved for public release via STI DAA 24818. See statement on title page.

Observations on Documentation Quality and Research Challenges

This chapter covers experiences and lessons learned from performing technical history research and knowledge capture. A considerable amount of research on Apollo and the Space Shuttle was conducted as part of the Orbital Space Plane (OSP) and Orion Programs. These experiences and lessons influenced the knowledge capture approach taken by United Space Alliance personnel when writing some of the reports discussed in this volume, as well as recommendations for knowledge capture given to the NASA/JSC Mission Operations Directorate at the end of the Shuttle Program.[1,2]

---

[1] Goodman, John, "Lessons Learned From Lessons Learned," presented at the USA Engineering Team Lessons Learned Summit With USA Constellation Personnel, San Antonio, TX, July 8, 2008.

[2] Goodman, John, "Thoughts on Knowledge Capture for Shuttle Retirement and Transition," presentation to the NASA/JSC Flight Design and Dynamics Division (DM) Issues Meeting, January 11, 2010.

Approved for public release via STI DAA 24818. See statement on title page.

Programmatic Histories Compared to Technical Documentation

Formal space program histories are usually written to address broad programmatic, policy, political, diplomatic, economic, cultural, and program management topics.[1] Identifying and understanding past program challenges and applying lessons learned at the programmatic management and agency policy level is just as important for the success of space flight programs as identifying challenges, and applying lessons learned and best practices at the technical level.[2,3] Other historical works are written for the general public (often with a human interest angle) to increase understanding of and appreciation for space flight technical and scientific accomplishments. These works are of value to historians, space policy specialists, and members of the general public that are space enthusiasts.

Traditional histories may also provide useful background information for engineers and managers developing vehicle and supporting systems and operations concepts (crew and ground activities including mission planning, hardware and software preparation, training, mission execution, and post flight activities). For example, the decision to adopt the Lunar Orbit Rendezvous (LOR) mission profile over other, higher risk profiles for Apollo was a key to meeting President Kennedy's goal of landing a crew on the Moon and returning them safely to Earth before 1970. The NASA History Monograph *Enchanted Rendezvous* describes the advocacy role played by John Houbolt during the lunar mission profile debates in 1961 and 1962 that led to the adoption of LOR. The monograph illustrates for future generations of engineers and managers the important role of advocacy and associated communication skills in getting a flight program to adopt the best and lowest risk technical solution.[4]

While traditional historical works are valuable for understanding the big picture of a flight program, they typically contain little technical detail that is useful for identifying technical lessons learned, how technical challenges were overcome, or rationale behind

---

[1] For an overview of the historiography and sub-genres of historical works concerning the U.S. space program see Siddiqi, Asif A., "American Space History: Legacies, Questions, and Opportunities for Future Research," chapter 14 in *Critical Issues in the History of Space Flight*, NASA SP-2006-4702, edited by Steven J. Dick and Roger D. Launius, NASA, Washington, DC, 2006.

[2] An excellent example is the National Polar Environment Satellite System (NPOESS), canceled in February of 2010 by the White House. See Hall, Tim, "NPOESS Lessons Evaluation, Executive Summary," Aerospace Corporation, December 1, 2010. At the time of publication available at the NOAA Office of Systems Development website, http://www.osd.noaa.gov/download/NPOESSAerospaceReport.pdf .

[3] For examples from a U.S. Air Force space perspective see:
Taverney, Thomas D., and James D. Rendleman, "Ten Rules for Common Sense Space Acquisition," *High Frontier*, Vol. 6, No. 1, November 2009, pages 53-65.
Rendleman, James D., and J. Walter Faulconer, "Escaping the Space Acquisition Death Spiral," Parts 1, 2, and 3, *High Frontier*, Vol. 7, No. 4, August 2011. *High Frontier* was published by the United States Air Force Space Command (http://www.afspc.af.mil/library/highfrontier journalarchive.asp).

[4] Hansen, James R., *Enchanted Rendezvous: John C. Houbolt and the Genesis of the Lunar-Orbit Rendezvous Concept*, Monograph in Aerospace History, No. 4, NASA, Washington, DC, 1995. The author, Dr. James Hansen, later wrote the authorized biography of Neil Armstrong, *First Man*.

Approved for public release via STI DAA 24818. See statement on title page.

vehicle systems requirements, system design, spacecraft and ground facility operation, and mission planning.[5] Furthermore, traditional histories are usually written with limitations on content driven by export control regulations, limited manuscript length, and limits on cost and schedule to ensure that the publisher makes a profit.[6, 7, 8]

---

[5] The NASA history of the Apollo/Soyuz Test Project (ASTP), published in 1978, was researched and written by the authors while embedded with ASTP personnel at the NASA Johnson Space Center. While they had to be selective in terms of level of detail and what to cover, they wrote about topics and discussions that might not have been covered in a more traditional history. For example, details of ASTP docking hardware development and Reaction Control System jet plume impingement were of interest to later engineers working on the shuttle missions to Mir and ISS. However, this non-traditional approach to writing history along with publication three years after the ASTP mission did not permit the authors to place ASTP in the broader historical context of later Cold War events and the eventual break-up of the Soviet Union that ended the Cold War. See Ezell, E. C., and L. N. Ezell, *The Partnership, A History of the Apollo-Soyuz Test Project*, NASA SP-4209, NASA, Washington, DC, 1978.

[6] An excellent example of a book length technical history is the 681 page tome by Dennis Jenkins on the X-15 program. This book contains a significant amount of technical detail, woven into a story along with the programmatic and historical aspects of the X-15 that would be useful resource for personnel working on future hypersonic flight programs. It bridges the gap between traditional programmatic histories and detailed technical reports. Unfortunately, books containing this level of detail are not available for the Apollo, Space Shuttle, and International Space Station Programs. See Jenkins, Dennis R., *X:15: Extending the Frontiers of Flight*, NASA SP-2007-562, NASA, Washington, DC, 2007.

[7] Another excellent example of a book length technical history of a flight program is the three volume work by Dr. William Corliss on the Pioneer 6, 7, 8, and 9 interplanetary probes of the 1960s.

Corliss, W. R., *The Interplanetary Pioneers. Volume 1: Summary*, NASA-SP-278, NASA, Washington, DC, 1972.

Corliss, W. R., *The Interplanetary Pioneers, Volume 2: System Design and Development*, NASA-SP-279, NASA, Washington, DC, 1972.

Corliss, W. R., *The Interplanetary Pioneers. Volume 3: Operations*, NASA-SP-280, NASA, Washington, DC, 1972.

[8] The NASA publication *Wings In Orbit* is an excellent overview of the accomplishments of the Shuttle Program. However, it is written for members of the general public interested in science and technology at a *Scientific American* reading level, rather than for technical personnel working on future flight programs. See Hale, Wayne (executive editor), *Wings In Orbit: Scientific and Engineering Legacies of the Space Shuttle*, NASA/SP-2010-3409, NASA, Washington, DC, 2011.

Approved for public release via STI DAA 24818. See statement on title page.

Relative Quality of Apollo and Space Shuttle Documentation

The best sources of Apollo and early shuttle information were detailed technical reports written by technical personnel at the time of a mission or when an analysis was conducted. The most informative reports were written using the assumption that a future reader might not have worked the mission. Reports written with complete sentences and paragraphs were more educational than presentations using bullet points or data and cryptic notes in spreadsheets. Apollo documentation was often better written and more informative than shuttle documentation. Most Apollo documentation consists of informal memos and formal reports written in complete sentences and paragraphs, while most shuttle documentation consisted of bullet charts and spreadsheets. Some personnel discovered that they acquired a better understanding of what happened during Apollo missions than during more recent shuttle missions.[1] Conference papers written during the Apollo Program were also informative.[2]

The Apollo Experience Reports were particularly helpful to Orion personnel researching Apollo.[3] In March of 1969, four months before the Apollo 11 lunar landing, Apollo Program management recognized the need to document the technical and management knowledge and experience of the Apollo Program. Direction was given in May of 1969 (two months before the first lunar landing) for reports to be written about each major spacecraft sub-system and major system.[4] These reports would outline Apollo system design concepts and recommend changes for future flight programs. The reports were to detail the underlying philosophical considerations behind systems design (why things were they way they were) and failed design approaches. In addition, the reports also covered project management, flight operations, mission planning, and crew training. An Apollo Experience Report Editorial Committee (AEREC) was established in December 1969. Due to the comprehensive coverage of the experience reports the Apollo Program decided in March of 1970 to cancel post-flight technical conferences similar to those conducted during the Mercury and Gemini Programs.[5,6,7] Eventually about 121 experience reports were published, with most approved for publication by 1973.[8] Reports

---

[1] The *NASA Mission Reports* series published by Apogee Books was particularly useful for Apollo research. These reports contained, for each mission, numerous and technically detailed press releases, post flight reports, and transcripts of the post-flight crew debriefings.

[2] The MIT Instrumentation Laboratory, later renamed the Charles Stark Draper Laboratory, published many informative papers during the Apollo Program concerning the primary guidance, navigation, and control system on the Command and Lunar Modules.

[3] Holman, D. N., "The Apollo Experience Reports," *20th International Technical Communications Conference*, Houston, TX, May 9-12, 1973, sponsored by the Society for Technical Communication.

[4] The first half of 1969 was arguably the busiest period in the history of NASA's human flight program.

[5] *Mercury Project Summary – Including Results Of The Fourth Manned Orbital Flight*, May 15-16, 1963, NASA SP-45, Washington, D.C., October, 1963.

[6] *Gemini Midprogram Conference Including Experiment Results*, NASA SP-121, NASA Manned Spacecraft Center, Houston, TX, February 23-25, 1966.

[7] *Gemini Summary Conference*, NASA SP-138, NASA Manned Spacecraft Center, Houston, TX, February 1-2, 1967.

[8] At the time of this publication the reports were available through the NASA Technical Reports Server website on the internet.

Approved for public release via STI DAA 24818. See statement on title page.

similar to *The Apollo Experience Reports* can enable personnel working in future spacecraft development programs to quickly identify challenges faced by previous programs, and determine if those challenges and associated technical approaches are applicable to a new development effort. Challenges that were not overcome in a previous program may be overcome in a future program due to advances in technology.

There are four reasons why Apollo documentation was of higher quality and more available than shuttle documentation. Many personnel that participated in Apollo development in the 1960s remained with the program into the early 1970s and were able to contribute to the research and writing of various types of knowledge capture documentation, such as *The Apollo Experience Reports*. On the other hand many 1970s shuttle development personnel left the Shuttle Program in the 1980s and 1990s and were not available to contribute to later shuttle knowledge capture efforts. The second reason concerned the much higher flight rate of the Space Shuttle as compared to the Apollo lunar missions. Shuttle personnel (and Mission Control, crew training, and flight planning personnel in particular) began working a future mission as soon as a mission was completed. The heavy flight-to-flight work load with little or no "down time" between mission completion and preparation for the next flight made it difficult to capture experiences and lessons learned through memos, formal reports, and presentations.[9] Third, the work culture at the time of the Apollo Program appears to have valued formal documentation more than the work culture in the shuttle era. Contractors appear to have been required by NASA to prepare formal reports more than during the Shuttle or even Orion Programs. The fourth reason, already mentioned, was that much Apollo documentation was informal memos and formal reports written using complete sentences and paragraphs while bullet charts were preferred during the Shuttle and Orion Programs.

Formal reports containing technical detail in historical context written by technical personnel for a technical audience were helpful for rapidly gaining insight into the requirements, systems design, mission activities and flight performance results from previous flight programs.[10,11] Such insight is required to determine if algorithms,

---

[9] The book *History of Space Shuttle Rendezvous* (JSC-63400, Revision 3) contains a chapter titled "STS-130 Mission to the ISS." It is a 28 page knowledge capture chapter detailing a typical shuttle mission to the ISS from a rendezvous and proximity operations perspective. The author performed much knowledge capture (email, Mission Control console notes, presentations, informal interviews) during the mission as he worked the flight. The chapter was researched and written during the two months following the February 2010 mission. Memories of crew members, Mission Control personnel, and the crew were fresh, resulting in excellent feedback to the author on the draft. Performing knowledge capture, research, and writing concurrent with and immediately after the events of interest resulted in a higher quality document than could have been created by an effort conducted many months or years after the flight.

[10] Excellent examples of informative technical reports with historical context from NASA's aeronautics research are:

Day, Richard E., *Coupling Dynamics in Aircraft: A Historical Perspective*, NASA-SP-532, NASA Dryden Flight Research Center, Edwards, California, March 1997.

Day, Richard E., *Energy Management of Manned Boost-Glide Vehicles: A Historical Perspective*, NASA/TP-2004-212037, NASA Dryden Flight Research Center, Edwards, California, May 2004.

Approved for public release via STI DAA 24818. See statement on title page.

requirements, lessons learned, and mission flight techniques from previous programs are applicable to a new development and flight program.

---

[11] The five volumes that detail the MIT Instrumentation Laboratory (later called the Charles Stark Draper Laboratory) guidance, navigation, control, and software contributions to the Apollo Program are also good examples.

MIT's Role in Project Apollo, Volume *1 Project Management, Systems Development, Abstracts and Bibliography*, Charles Stark Draper Laboratory, R-700-VOL-1, October 1971.

MIT's Role in Project Apollo, Volume 2: Optical, Radar, and Candidate Subsystems, Charles Stark Draper Laboratory, R-700-VOL-2, NASA-CR-141898, March 1, 1972.

Hall, E. C., *MIT's Role in Project Apollo, Volume 3: Computer Subsystem*, Charles Stark Draper Laboratory, R-700-VOL-3, NASA-CR-151191, August 1, 1972.

MIT's Role in Project Apollo, Volume 4: Inertial Subsystem, Charles Stark Draper Laboratory, R-700-VOL-4, NASA-CR-141897, April 1, 1972.

Johnson, M. S., and D. R. Giller, *MIT's Role in Project Apollo, Volume 5: The Software Effort*, Charles Stark Draper Laboratory, R-700-VOL-5, NASA-CR-140339, July 1, 1971.

Approved for public release via STI DAA 24818. See statement on title page.

Apollo and Space Shuttle Research Challenges

Extensive bibliographies and references enable engineers to quickly identify conference papers or internal documentation of interest that can be obtained through NASA center libraries. Such reports can save considerable time since cost and schedule constrained vehicle development places a heavy workload on technical and management personnel. One example is a JSC document providing abstracts of 37 technical reports written by Dr. William M. Lear from 1973 to 1995.[1]

Some engineers may have difficulty understanding legacy reports if they are not familiar with the details of the flight program in question. This makes it difficult to determine if the technical solutions, lessons learned, and challenges of a legacy program are relevant to a new program. Personnel who grew up with the internet may not be familiar with the contents of paper libraries or papers that were published before the internet was created.

Most technical and management personnel do not have an aptitude for extensive and time consuming research using primary and secondary source materials. Librarians and archivists can speed-up research by locating primary and source materials for engineering personnel. However, in some cases more primary source materials were collected than engineers had time to examine.

Apollo and shuttle research experience showed that contractually deliverable documents are often preserved in government agency and corporate libraries. However, these documents do not contain the rationale used to create the contractual deliverables. This decreases the usefulness of the documentation to future engineers. Lower level documentation that provides insight into contractually deliverable documentation should be preserved.

Many organizational libraries containing Apollo and shuttle documentation have disappeared for two reasons. First, the need for floor or filing cabinet space resulted in libraries being recycled or sent to archive (and retrieving archived materials is not always easy). Second, personnel who owned the libraries and were knowledgeable of and appreciated the contents retired or moved to other jobs. They were no longer present to champion the need for the documentation or explain why the organizational libraries were of value.

---

[1] Dr. Lear made significant contributions in the areas of Kalman filtering, navigation, and dynamics during a 40 year career in the aerospace industry with TRW and Draper Laboratory. He developed the Apollo Powered Flight Processor (PFP) that processed data from three or four ground tracking stations to determine the LM navigation state during Lunar Module powered descent and powered ascent. The PFP was used by Mission Control to assess the health of the LM primary and backup computers and later formed the basis for Mission Control ground navigation during shuttle entry. See Lear, William M., *Abstracts of Technical Reports Written By William M. Lear, Ph.D.*, JSC-27368, NASA/JSC Aeroscience and Flight Mechanics Division, January 1996.
Approved for public release via STI DAA 24818. See statement on title page.

Much important and informative documentation was not in formal libraries, but was preserved by individuals on an ad hoc basis in desks, filing cabinets, and on computer hard drives.[2, 3] Important documentation can be lost during government agency and corporate reorganizations and contract transitions. However, personnel tend to retain the documentation that was most important to them.

Organizations may have difficulty determining what documentation was sent to an archive in the past. The cost of retrieving documentation from off-site archives can deter organizations from making use of archived resources. Furthermore, archived resources may not be adequately indexed. This forces the retrieval of large amounts of documentation (tens of linear feet) that require manual and time consuming examination to find relevant documents. Documentation often arrives at archives with little or no information (such as abstracts) that would enable archivists to index it, catalog it, and assign keywords. While archivists were very helpful, they are not subject matter experts and they depend on the organizations donating the documentation to provide descriptions, key words, and historical context.

Although a few books cover shuttle development, an easily accessible timeline of shuttle development milestones from contract award in 1972 to the first shuttle mission STS-1 (April 1981) and accompanying supporting documentation did not appear to exist.[4] Such information may exist in primary or secondary documents preserved in archives. Retired personnel remembered participating in programmatic reviews but none of the personnel interviewed could provide a complete picture of the number and types of reviews or the issues discussed and closed.[5] Requirements rationale and descriptions of the mission planning process were difficult or impossible to find. Rationale behind contingency

---

[2] At the end of the Shuttle Program shuttle rendezvous crew trainer Jorge Frank of United Space Alliance conducted extensive Apollo and shuttle research and knowledge capture. He observed that it was difficult to find informal shuttle memos and presentations dating from approximately 1983 to 1993. Frank believed that during this period much informal documentation was stored on floppy disks and desktop computers, as opposed to the traditional and exclusive use of paper. By 1993 desktop computers were networked to external network hard drives, increasing the probability of preservation.

[3] For a summary of experiences and lessons from an end of Shuttle Program effort to collect and archive documentation, see: Goodman, Doug, Chip Shepherd, and Al Godbout, "NASA Space Shuttle Program at Johnson Space Center Engineering Directorate Records Archival: Lessons Learned," *AIAA SPACE 2010 Conference and Exposition*, Anaheim, CA, August 30 - September 2, 2010.

[4] Excellent overviews of Space Shuttle development are:

Heppenheimer, T. A., *Space Shuttle Decision*, 1965-1972 (History of the Space Shuttle, Volume 1), Smithsonian Books, Washington, DC, 2002.

Heppenheimer, T. A., *Development of the Space Shuttle,* 1972-1981 (History of the Space Shuttle, Volume 2), Smithsonian Books, Washington, DC, 2002.

Jenkins, D. R., *Space Shuttle – The History of the National Space Transportation System – The First 100 Missions*, Specialty Press Publishers, North Branch, MN, 2001.

[5] This is not unusual. In 1971 *Stages to Saturn* author Roger Bilstein commented during an interview of NASA Marshall Space Flight Center Aero-Astrodynamics Laboratory Director Ernst D. Geissler that he was having difficulty determining the dates of major decisions made concerning the design of the Saturn launch vehicles, even though it had only been 10 years since the decisions were made. See Ernst D. Geissler interview, NASA Marshall Space Flight Center, September 7, 1971, located in the Stages to Saturn History Project, Oral History Collections, University of Alabama Huntsville Library. Bilstein, Roger E. *Stages to Saturn: A Technological History of the Apollo/Saturn Launch Vehicles*, NASA SP-4206, NASA, Washington, DC, 1980 and 1996. Reprinted by The University Press of Florida, 2003.

Approved for public release via STI DAA 24818. See statement on title page.

procedure development and software changes was difficult to locate. Flight Software Change Requests (CRs) usually survive without the supporting documentation that details CR rationale, development, and testing. Source code can be an important reference for algorithms even if computers can no longer execute the source code.

In 2009 an Integrated Design Assessment Team (IDAT) was formed to study options for the design of the Orion parachute system. The team sought input from Apollo veterans and surviving documentation. Locating relevant documentation was challenging, but a large amount of documentation was eventually obtained. Sources were NASA databases, Apollo veterans, and a Northrup archive. Documents from the Northrup archive were scanned to obtain electronic copies. The documents were thorough, well written, and important sources of information. Input from Apollo veterans was valuable during the IDAT activity and after.

An example of adapting a legacy algorithm versus developing a new one concerns Orion entry guidance. Apollo had a skip entry requirement to provide landing options in the event of undesirable weather conditions at the primary landing site.[6] Although Apollo skip entry guidance was certified, it was never flown, due to the difficulty in flying manual skip entries in case of a computer failure, and due to navigation and guidance inaccuracies.[7] Orion also had a skip entry requirement to permit landing at the same point on Earth for lunar landings at latitudes ranging from equatorial to polar. The final phase of Apollo entry guidance was adopted for Orion due to its excellent performance. However, greater Orion computer capacity permitted Orion personnel developed a new skip entry guidance algorithm, used at altitudes higher than final phase guidance, with better performance than the legacy Apollo skip guidance algorithm.[8,9,10] Analysis of the skip guidance problem required extensive searches to locate Apollo documentation.

Much Apollo and early shuttle documentation survived but it was difficult for researchers to determine why reports were written or what was done based on the reports. Interviews of personnel that possessed a sharp memory were useful for understanding the historical and programmatic context of documentation. These personnel were able to fill in the gaps of knowledge of Apollo and early shuttle technical history.

Some oral history transcripts were also examined. The challenge of using interviews and oral history transcripts dated long after the events in question is that memories fade,

---

[6] Burton, John K., *Introduction to Apollo Entry Guidance and Flight Performance*, JSC-35020, NASA JSC Flight Design and Dynamics Division, July 2009. Discussed in this document.

[7] Graves, C. A., and J. C. Harpold, *Apollo Experience Report: Mission Planning For Apollo Entry*, NASA-TN-D-6725, NASA Manned Spacecraft Center, March 1972.

[8] Tigges, Michael A., Timothy Crull, Jeremy Rea and Wyatt Johnson, "Numerical Skip-Entry Guidance," *Guidance And Control 2007, Advances In The Astronautical Sciences*, Volume 128, Univelt, Inc., San Diego, CA, 2007.

[9] Bairstow, Sarah, and Gregory Barton, "Orion Reentry Guidance with Extended Range Capability Using PredGuid," *AIAA Guidance, Navigation and Control Conference and Exhibit*, Hilton Head, South Carolina, August 20-23, 2007.

[10] Rea, Jeremy, and Zachary Putnam, "A Comparison of Two Orion Skip Entry Guidance Algorithms," *AIAA Guidance, Navigation and Control Conference and Exhibit*, Hilton Head, South Carolina, August 20-23, 2007.

Approved for public release via STI DAA 24818. See statement on title page.

technical details are forgotten, and stories change over time. Oral history is most accurate when performed during or just after the events of interest. Oral history is particularly valuable for researching the programmatic, political, policy, and cultural aspects of space flight. Audio or video interviews may be the only practical way of capturing experiences and knowledge of busy technical personnel and program leaders who may not possess written communications skills.

Knowledge Capture via Software Documentation

When the Space Shuttle was designed and built in the 1970s it had the most sophisticated software and avionics architecture ever designed for an aerospace vehicle up to that time.[1] Flight performance of both the Primary Avionics Software System (PASS) and Backup Flight System (BFS) was been outstanding over the life of the program.[2,3]

Many factors could be cited for the success of the PASS and BFS flight software. Among them are robust development, testing, configuration control, issue investigation, and issue resolution processes.[4] Multiple NASA and contractor organizations participated in various phases of requirements development, test scenario development, test data evaluation, and flight software issue resolution. Extensive interaction of Shuttle Program personnel with skilled PASS (IBM, then Loral, then Lockheed Martin, and finally United Space Alliance) and BFS (Rockwell International, later Boeing) software specialists occurred throughout the ~35 year life of the shuttle software effort.

Another success factor was the insight that NASA and contractor personnel had into the PASS and BFS equation level requirements and software functionality. The primary documents that provided this insight was the configuration controlled PASS Functional Subsystem Software Requirements (FSSR, also known as Level C) and the BFS Program Requirements Document (PRD) documents. The shuttle FSSRs and PRDs provide detailed and accurate equation level design and logic information that was written by engineers and understandable by engineers.[5] These documents served as the authoritative, configuration controlled source of software algorithm insight for personnel across the Shuttle Program. The FSSR and PRD volumes ensured that NASA and contractor personnel in various program elements could obtain the same level of software insight.

The original FSSR and PRD volumes were written in the mid and late 1970s. The volumes underwent extensive reviews by Shuttle Program personnel that did not write them. Reviewers conducted thorough inspection of the requirements, equations, and interfaces. These reviews and subsequent corrections by a large number of personnel enabled problems to be identified and corrected before flight software was written. The

---

[1] Hanaway, John F., and Robert W. Moorehead, *Space Shuttle Avionics System*, NASA SP-504, National Aeronautics and Space Administration, 1989.

[2] Shuttle development veterans from the 1970s told the author that during the shuttle design phase people thought that if a vehicle was ever lost, it would most likely be due to software or some other avionics failure. Such a loss did not occurr. *Challenger* and *Columbia* were lost due to hardware failures.

[3] Zimpfer, Douglas, Phil Hattis, John Ruppert, and Don Gavert, "Space Shuttle GN&C Development History and Evolution," *AIAA SPACE 2011 Conference & Exposition*, AIAA, Long Beach, CA, September 27-29, 2011.

[4] Hickey, Christopher J., James B. Loveall, James K. Orr, and Andrew L. Klausman, "The Legacy of Space Shuttle Flight Software," *AIAA SPACE 2011 Conference & Exposition*, AIAA, Long Beach, CA, September 27-29, 2011.

[5] During the Apollo Program the equivalent documents for the Lunar Module and Command/Service Module software were the Guidance System Operations Plan (GSOP) volumes.

Approved for public release via STI DAA 24818. See statement on title page.

reviews also enabled many personnel, including software developers, to become familiar with the detailed software requirements and underlying theory of the algorithms.[6, 7]

Research into flight software functionality was a continuous activity conducted by both experienced and new personnel in numerous Shuttle Program disciplines over the 30 year flight program. There were many factors that drove this research – crew and ground procedure development, vehicle flight and ground test performance investigation, development of training documentation for the crew and other program personnel (analysts, flight controllers, etc.), simulator and mission planning tool development, identification and development of software upgrades, risk assessment, mission planning, trajectory design, etc. For example, the shuttle training facilities used the same flight software that flew on the shuttles. Detailed information on how the flight software worked was required by personnel who maintained the various models and other supporting software in the crew training simulators.

FSSRs and PRDs were the primary source of as-coded documentation for PASS and BFS software. FSSRs and PRDs, while not perfect, were invaluable aids to understanding PASS and BFS functionality since they were written by engineers, not software specialists. There was a large community of non-software specialists that regularly used shuttle FSSR and PRD requirements as a part of their work. FSSRs and PRDs enabled non-software specialists to acquire accurate and timely insight into how the PASS and BFS software worked and should perform. The existence of the FSSRs and PRDs saved the Shuttle Program a considerable amount of time and money. Furthermore, the easy access to accurate pseudo-code, interface tables, and paragraphs explaining functionality increased the knowledge and insight of Shuttle Program personnel. This in turn enhanced mission success and safety of flight.

The shuttle PASS FSSR contained useful flowcharts. However, use of flow charts did not provide sufficient insight into all aspects of software functionality and logic needed by engineers outside the software process. FSSR paragraphs detailing software requirements, interfaces, and functionality were invaluable. Personnel needed accurate documentation of not only the as-coded equations, but the interface logic that tied the large number of software modules together. The FSSRs provided this information. Unfortunately, the flight control FSSRs contained block diagrams but no paragraphs to explain the flight control logic. This made flight control software research more difficult.

Not all flight software questions could be resolved by the examination of FSSRs and PRDs by engineers external to the flight software development and maintenance organizations. Some issues required the involvement of PASS and BFS software

---

[6] Shuttle software development organizations included many personnel with engineering, physics, and applied math backgrounds. These personnel understood the guidance, navigation, and control theory underlying the software they were implementing.

[7] For a discussion of the value of FSSRs during shuttle flight software development in the late 1970s see the comments by shuttle guidance sub-system manager Aldo Bordano in *Compilation of Powered Explicit Guidance Papers*, NASA/JSC Flight Dynamics Division, JSC-64694, February 2010, pages 40 to 42.
Approved for public release via STI DAA 24818. See statement on title page.

specialists. These issues included questionable software performance that was formally documented in discrepancy reports as well as proposals for software changes.

Frequently other questions arose did not warrant writing a flight software discrepancy report that would kick-off a formal investigation by the PASS or BFS development organizations. Software specialists with access to HAL source code, additional documentation beyond the FSSRs, and Shuttle General Purpose Computers (GPCs) for running tests were able to provide answers and insight beyond what many engineers could acquire simply by studying the FSSR or PRD.

Software development, maintenance, and flight-to-flight reconfiguration personnel had access to the following resources and documentation during software performance and functionality investigations:

    1) Shuttle computers running flight software.

    2) HAL source code.

    3) FSSRs and PRDs – Pseudo-code and detailed equation level requirements.

    4) Detailed Design Specification (DDS) – How the FSSRs and PRDs requirements were coded.

    5) Maintenance Specification – Higher level description of flight software architecture.

    6) Equations Derivations – If they could be found. They usually could not be located.

However, there was a limit to how much investigation that members of the PASS and BFS software development and maintenance organizations could perform for other personnel in the Shuttle Program. Such investigations could be performed only on a non-interference basis, unless of course a performance issue arose that was documented through a flight software discrepancy report or some other formal documentation. The existence of the FSSRs and PRDs enabled personnel outside the flight software community to conduct daily research and obtain answers without consulting the flight software organizations. If the FSSRs and PRDs did not exist, the flight software community would not have been able address the large number of questions and need for information that arises as a part of mission preparation and execution.

The shuttle PASS and BFS flight software was high quality source code and was the product of a strict and robust development process. While extensive comments were included in the source code, the comments were not enough to enable shuttle personnel to understand how it functioned, investigate anomalous performance, or develop software changes. Commented source code was not a substitute for the configuration controlled FSSR and PRD volumes. Source code was not an effective or efficient way for engineers who were not software specialists to study software functionality. While the PASS Detailed Design Specification and the Maintenance Specification were important

Approved for public release via STI DAA 24818. See statement on title page.

documents written and owned by the PASS flight software organization, they did not supply the same level of equation level insight as the FSSR. The FSSR was more useful to non-software specialists and easier to understand than the PASS HAL source code, Detailed Design Specification, and the Maintenance Specification.

On-board software for the International Space Station (ISS) was developed under the assumption that detailed FSSR type requirements documents were not required. As a result software insight was a continual challenge during ISS flight operations.[8] The ISS flight software was written in Ada (hand coded) and MatrixX. ISS Mission Control organizations did not have enough personnel to employee software specialists devoted to Ada and MatrixX analysis. MatrixX source code was difficult to understand unless one worked with it on a regular basis. ISS flight controllers obtained flight software functionality and performance insight through the following means:

1) Real time software tests run on the training simulators or the ISS Software Verification Facility (SVF).[9]

2) Technical Description Document covering the mathematics in the attitude control software, last updated in 1996.

3) Attitude determination mathematics added by Boeing engineers to the Software Requirements Summary (SRS) as an ad hoc form of knowledge capture.

4) Social networking with McDonnell Douglas/Boeing ISS engineers and software developers.

The effectiveness of gaining software insight through social networking depended on establishing and maintaining long-term relationships between ISS flight controllers and ISS engineering and software personnel. The most effective relationships were ones that dated back to college days. Not all ISS Mission Control disciplines were able to effectively establish such relationships. The lack of detailed software documentation also made it difficult to train new ISS engineering personnel and investigate ISS performance anomalies.

Shuttle flight software documentation provided invaluable and necessary insight into the safety critical software that was maintained by multiple generations of personnel over the 30 year flight phase of the Shuttle Program. The difficulty encountered by ISS Mission Control personnel trying to understand the ISS software points to the value of such documentation. Future flight programs should possess flight software documentation that

---

[8] Email exchange between ISS Flight Director Michael Lammers and John Goodman dated November 23, 2009. Before becoming an ISS Mission Control Flight Director Mr. Lammers spent 10 years as an ISS flight controller specializing in flight software for Command and Control (C&C) and Guidance, Navigation, and flight Control (GNC).

[9] The SVF was part of the Software Development and Integration Laboratory (SDIL), located at the NASA/JSC Sonny Carter Training Facility (SCTF), Building 920A. It was the ISS version of the Shuttle Avionics Integration Laboratory (SAIL) in NASA/JSC Building 16.

Approved for public release via STI DAA 24818. See statement on title page.

is, at a minimum, equivalent to that in the Shuttle Program in terms of knowledge capture, readability, coverage of detail, and accessibility to personnel across a flight program.

Knowledge capture during software development may be complicated by the adoption of a "software is self documenting" philosophy similar to that of the International Space Station. Detailed software requirements documents may not be written in order to save development time and money. However, this can limit the number of personnel that fully understand software functionality and theory to those that coded the software. This may make it difficult for personnel not involved in software development to perform independent analysis and verification of the software. Furthermore, this approach can result in software reviews that focus on high level software capabilities, rather than detailed equation level reviews of algorithms and interface requirements that occurred during the Apollo and Space Shuttle Programs.

Specific requirements and contractual deliverables must exist for commenting source code, creating software insight memos, developing equation level requirements documents, and writing software design description documents. Otherwise, personnel will not perform these activities due to the heavy development environment workload and fast-paced schedule. Lack of such documentation will increase life cycle costs during mission planning, ground and on-board procedure development, software maintenance, and mission execution once the flight phase of a program begins. Thorough understanding of software design, functionality, performance, and the theoretical basis of algorithms is an essential component of a good flight safety culture.

Orion Knowledge Capture Challenges

Much vehicle design work conducted from Orion contract award (August 31, 2006) through the first flight of Orion will be of interest over the life of the Multi-Purpose Crew Vehicle Program. Important Orion design information such as requirements rationale, operations concepts, and technical development history resided in many formats (presentations, emails, spreadsheets, informal memos, formal reports, status reports, meeting minutes), both electronic and hardcopy. However, the budget, schedule, and employee work load constraints made it difficult to capture documentation in a manner that would provide historical context and ease of access to future engineers.

The priority during vehicle development is to meet cost, schedule, and delivery (hardware, software, documentation) requirements, and keep the program sold and in business. Knowledge capture may be seen as not contributing to successfully meeting these requirements and is therefore an optional activity. Budget constraints, a fast paced development schedule, and heavy work load can prevent management from dedicating subject matter experts to knowledge capture, or creating knowledge capture tasks.

Concerned personnel performed knowledge capture as a side activity on a non-interference basis with respect to regular work. These personnel had to be proactive and investigative to get access to informative documentation (emails, memos, presentations, etc.) and key personnel. Not all of the primary sources needed to assemble a technical development history could be obtained and preserved during the normal course of business. Fast paced vehicle development with continually evolving vehicle requirements and resource (schedule, time, and available personnel) constraints made it difficult for personnel to create detailed technical histories of sub-systems development.

Email was an important means of communications among development team members that were geographically separated. However, the large amount of email exchanged between team members, and NASA and contractor management in particular, was overwhelming. More email was generated on a daily basis than personnel could read and respond to in a timely manner. Limited use of email and greater use of face-to-face conversations, phone conversations, and informal memos was necessary to ensure effective communication in a timely manner. However, conversations and informal memos did not communicate with as wide an audience as emails. The necessary reduction in email traffic made it more difficult for some personnel to stay informed in the fast paced development environment. This made performing knowledge capture more difficult.

Summaries of recent decisions and updates on the status of tasks given by team leaders in meetings were useful for keeping personnel up to date, particularly in light of the reduction in email traffic. This information was also documented in meeting minutes. Detailed meeting minutes were written and published for meetings of the Rendezvous, Proximity Operations, and Docking Working Group (RPOD WG) and the Guidance and

Targeting Working Group (GTWG). These minutes will enable future Orion personnel to understand the context of presentation charts, the discussions that occurred during the presentations, and decisions made or actions taken based on the presentations.

In May of 2008 an independent panel of seven spacecraft guidance, navigation, and control (GNC) specialists reviewed the on-orbit GNC software and trajectory designs developed by the Orion Flight Dynamics On-Orbit Mode Team.[1] Digital audio recordings were made of each presentation to enable future Orion personnel to understand the development rationale behind the on-orbit GNC system.[2] A total of 27.6 hours of recording was performed. A 99 page listener's guide was written that correlated audio file elapsed time to each page in each presentation.[3]

NASA leadership of the Orion On-Orbit Mode Team encouraged NASA and contractor personnel to document GNC analysis and software algorithms in formal Flight Dynamics team memos.[4] A memo format and numbering system was implemented and a folder in an electronic repository for Flight Dynamics memos was created. Bullet charts were not acceptable for formal documentation of analysis, requirements rationale, and theory underlying software algorithms. Conference papers were also written during Orion development that performed high level knowledge capture.[5]

Like Apollo and shuttle documentation, most Orion documentation provided little or no historical context such as why the document or presentation was written and what actions were taken. Presentations are typically written for a specific meeting with a specific

---

[1] Mode Team – Multiple Organization Design Environment Team

[2] Digital audio recordings were made since the GNC specialist that conducted this task had audio recording experience and did not possess video equipment. A tight schedule and limited budget prevented the team from using professional video personnel and equipment.

[3] Goodman, John, *Listener's Guide to Orion Orbit GNC Peer Review Recordings*, United Space Alliance, May 27, 2008.

[4] This was also known as the Orbit Mode Team. One example of such a memo is:

D'Souza, Christopher, "A Primer on the Orion Absolute Navigation UDU Filter," EG Technical Brief, EG-DIV-11-24, NASA/JSC Engineering Directorate, February 28, 2011.

[5] Many Orion and Ares I guidance, navigation, control, and trajectory papers were presented at AIAA and AAS conferences. Six examples are:

Brazzel, Jack P., Chris D'Souza, Peter T. Spehar, Fred D. Clark, Chad Hanak and Tim Crain, "NASA CEV Rendezvous Proximity Operations and Docking GN&C Analysis," *Guidance and Control 2007, Advances in the Astronautical Sciences*, Volume 128, Univelt, Inc., San Diego, CA, 2008, pages 607-630.

Crain, T., M. Begley, M. Jackson, and J. Broome, "GN&C System Design in a Mass Reduction Exercise," *Guidance and Control 2008, Advances in the Astronautical Sciences*, Volume 131, Univelt, Inc., San Diego, CA, 2008, pages 407-424.

Dukeman, Greg, "Rapid Trajectory Optimization for the ARES I Launch Vehicle," *AIAA Guidance, Navigation and Control Conference and Exhibit*, Honolulu, Hawaii, August 18-21, 2008.

Chambers, Robert, "Seven Vehicles in One: Orion GN&C," *Proceedings of the AIAA SPACE 2008 Conference and Exposition*, San Diego, California, September 9-11, 2008.

Marchand, Belinda, Michael Weeks, Chad Smith, and Sara Scarritt, "Onboard Autonomous Targeting for the Trans-Earth Phase of Orion," *AIAA Journal of Guidance, Control, and Dynamics*, Vol. 33, No. 3, May-June 2010, pages 943-956.

Jackson, Mark, "Orion Flight Performance Design Trades," *AIAA Guidance, Navigation, and Control Conference*, Toronto, Ontario, August 2-5, 2010.

Approved for public release via STI DAA 24818. See statement on title page.

objective, not to provide background information and context (historical, technical, programmatic) to future researchers. Status reports should reference presentations and reports so that future readers will understand the context and significance of them. Status reports and meeting minutes should be preserved along with supporting documentation. Information harvested from status reports, emails, and presentations can be combined into sub-system development timelines or informal memos.

The significant amount of Orion documentation stored electronically on various computer networks provides the possibility that more Orion development documentation could be preserved than Apollo and shuttle development documentation. Electronic document repositories are useful for capturing presentations, informal memos, and formal reports during vehicle development. Furthermore, some electronic repositories require purchasing user licenses. Budget limits and organizational boundaries within a flight program can prevent access by personnel with a legitimate business need for the documentation. Organization of documentation (such as folder hierarchy and names) within a repository can either make it difficult or easy to find documents of interest. Electronic storage can eliminate the floor and filing cabinet space issue encountered with shuttle paper documentation. However, computer memory may become an issue. If electronic archives are to be preserved technical or management personnel could be designated as archive owners. This is not a day-to-day librarian function, but the owner could ensure that the existence of, contents of, and value of the archives remain within the corporate memory of government agency and contractor personnel. This function would include working with the information technology personnel to ensure that changes in operating systems and computer applications do not compromise the ability to access the documentation. It is difficult to predict what future changes could occur in software applications and operating systems.

A proven algorithm from a past flight program can be implemented in new software incorrectly by engineers that are not familiar with its underlying theory and operation. Even proven algorithms that have been in use for decades need to be thoroughly documented to lower technical risk during future adaptation and re-use. Legacy Apollo and Space Shuttle documentation was useful for educating Orion personnel on the underlying theory of heritage algorithms used by Orion.

Some Apollo and Space Shuttle memos contained equations providing important theoretical insight or algorithms that could be incorporated in analysis software. However, some memos did not present derivations of important equations or references to where derivations could be found. Attempts to reverse engineer derivations are time consuming and not always successful. An inability to verify a derivation and identify and understand any assumptions made will lead engineers to not use the equation or algorithm in question, even if there is potential for cost and schedule savings by re-using it.

NASA personnel overseeing Orion GNC algorithm development encouraged engineers to write formal memos detailing algorithms developed for the Orion on-board software.

Many of these formal memos contained appendices with derivations of the algorithms.[6] This ensures that future engineers will have a good understanding of the algorithm and makes future reuse of the algorithm more likely.

A second example of documenting derivations and underlying theory concerns a technical note published in a journal in 1984 by a former NASA Ames engineer. The note described a new algorithm for solving Lambert's problem for orbital burn targeting.[7] The algorithm used an innovative new form of Kepler's equation that was presented without derivation. However, the references provided for the new form of Kepler's equation were privately published memos that readers of the journal would most likely not be able to locate. In 1996 the author published a second technical note in the same journal that presented the complete derivation of the new form of Kepler's equation.[8] The 1996 technical note also provided useful theoretical insight needed to fully understand the algorithm presented in the 1984 technical note. The author stated in the 1996 note that:

> *"Although the transformed equation has already been presented in essentially its final form and used as a basis for development of a new and efficient p-iteration algorithm [8], as noted above, its derivation has not been fully documented in an archival journal."*

This action by the author makes it more likely that the algorithm will be used in the future and that future engineers will benefit from the theoretical insight provided by the derivation and rationale behind the new algorithm.

---

[6] One example is D'Souza, Christopher, "A Primer on the Orion Absolute Navigation UDU Filter," EG Technical Brief, EG-DIV-11-24, NASA/JSC Engineering Directorate, February 28, 2011.

[7] Boltz, Frederick W., "Second-Order p-iterative Solution of the Lambert/Gauss Problem," *Journal of the Astronautical Sciences*, Vol. 32, No. 4, October-December 1984, pages 475-485.

[8] Boltz, Frederick W., "General Transformation of Kepler's Equation for p-Iterative Solution of the Lambert/Gauss Problem," *Journal of the Astronautical Sciences*, Vol. 44, No. 1, January-March 1996, pages 21-37.

Approved for public release via STI DAA 24818. See statement on title page.

Space Shuttle Knowledge Capture Approach

This chapter contains discussions of the knowledge capture approach taken during the research and writing of the reports detailed in this document.

The Knowledge Capture and Management Process

The knowledge capture and management process can be divided into four steps executed in the following order:[1]

1. People
2. Content
3. Process
4. Information Technology (computers, software)

Many corporations and government agencies begin implementing knowledge capture and management starting with step 4, the identification, procurement, and integration of computer hardware and software. A process (step 3) is then created that fits with the information technology. Then content to place in the knowledge management process is defined (step 2). Finally, personnel are identified that can provide and create knowledge capture content for the knowledge management process (step 1). This reverse approach (steps 4, 3, 2, 1) may be due to a false perception that subject matter experts cannot be identified and knowledge capture cannot be performed unless optimal corporation or agency wide knowledge management processes are in place using the latest information technology.

This reverse approach results in little overall benefit to technical and lower level management personnel. Optimal knowledge management process creation and implementation (step 3) and information technology (step 4) require resolving corporation or agency wide policy, process, budget, and information technology issues. This can require a considerable amount of time and budget. Establishing a large corporate or agency wide integrated process may not accomplish much knowledge capture in a cost and schedule constrained environment. Corporations and government agencies typically have far more personnel skilled at process creation (step 3), process management, and information technology (step 4) than they do technical subject matter experts (step 1) that possess the written, verbal, and graphic communications skills need to effectively perform knowledge capture (step 2).

An initial focus on the processes of sharing and retrieval, corporate or agency level processes and policy issues, benchmarking, trade studies, or abstract knowledge management theory that is difficult to understand and apply limits that actual knowledge capture and management that is performed by technical personnel. The difficulties encountered with a reverse approach can make knowledge capture and management harder than it really is. Significant expenditure of time and resources on steps 3 and 4 without results visible to and useful by lower level management and technical personnel could discourage personnel from initiating and participating in knowledge capture and management efforts.

---

[1] Lee, Jim, "Using Knowledge Management to Improve Project Success," presentation to the NASA Johnson Space Center Technical Speakers Forum, July 7, 2010. At the time of the presentation Jim Lee was Knowledge Management Practice Leader at APQC (the American Productivity & Quality Center).
Approved for public release via STI DAA 24818. See statement on title page.

The focus of personnel that researched and wrote the reports listed in this document was on steps 1 and 2. Step 1, people, involved identifying subject matter experts that possessed the verbal, writing, and graphic communications skills needed to perform knowledge capture. These steps can be performed by subject matter experts and their immediate management using existing processes and information technology resources without waiting for corporate or agency level process, policy, and information technology issues to be resolved. Subject matter experts and lower level management have little ability to influence or resolve the budget, policy, and process issues involved with steps 3 and 4. Identification of subject matter experts with knowledge capture skills and creation of knowledge capture documents and other media is critical in a cost and schedule constrained environment. Steps 3 (process creation) and 4 (information technology) could be worked later. Authors of the reports detailed in this document and other United Space Alliance personnel devoted time and resources to actually doing knowledge capture rather than studying knowledge management theory, benchmarking knowledge management processes, or performing information technology trade studies.

While a significant amount of knowledge capture (steps 1 and 2) can be performed at the lower levels of corporations and government agencies, visible leadership and support from senior management is needed to ensure the success and continuity of such efforts, particularly for the creation and implementation of knowledge management processes (step 3) and identification, procurement, and integration of information technology (step 4).

Leading and Managing People is the Key to Effective Knowledge Capture

Performing effective knowledge capture is not an information technology problem, but a people skills and work culture challenge. The most effective knowledge capture is accomplished by subject matter experts who possess visual, verbal, inquiry, and written communication skills. Unfortunately, not all subject matter experts possess these skills.

There is a perception that with optimal taxonomies and the latest search engine technology one will obtain effortless and instantaneous access to answers for whatever questions arise. This is an unrealistic expectation. In reality, while taxonomies and advanced search engines are helpful, they are not a replacement for intellectually curious personnel who enjoy learning and performing research. Technical personnel need to spend time learning what information sources are available and what they contain. The better understanding one has of sources and a topic the easier it is to find applicable information. Searching for answers is like detective work. The more experience one has the better they are at it. Even with the Internet, social networking and search engines, research still requires time and work. And developing research skills takes time.

Good engineers and managers are interested in developing the technical, leadership, and communication skills of personnel and avoiding single point failures in terms of key subject matter experts. Organizations will always have an attrition rate, personnel will change and documentation may be lost during re-organizations, and all programs come to an end at some point.

The best forms of knowledge capture are well-written and informative memos, presentations, formal reports, meeting minutes, and status reports that provide technical information, decision rationale, lessons learned, and requirements in the context of spacecraft system or overall program history.

Below are four key components of knowledge capture and management. All involve the leadership and management of people:

> **Discipline** – Place key knowledge and lessons in status reports, meeting minutes, presentations, procedures, training materials, memos, formal reports, etc. Share verbally in various forums. Refer to presentations, memos, and formal reports in status reports so that the significance of the documentation will be understood years later.
>
> **Communication Skills** – Interviewing, discussion, and writing skills are required to effectively research and communicate lessons learned and technical history. Communication skills are just as important as technical skills such as deriving equations.
>
> **People Management** – Identify subject matter experts with good communication skills and use them in knowledge capture and

Approved for public release via STI DAA 24818. See statement on title page.

management efforts. People who have observed and experienced the negative impact of corporate knowledge loss or worked on challenging projects are more motivated to participate in such efforts, document lessons learned and experiences, and mentor other employees.

Leadership – All levels of management should encourage and reward employees who identify and document lessons and preserve knowledge. Stress the importance of developing and demonstrating good communication skills.

Previous generations of engineers (before 1980) understood theory well, but did not have fingertip access to powerful computing resources. The use of slide rules and nomography forced engineers to be proficient at devising analytical and empirical solutions. This was very important for creating software that would run in the low capacity computers of the 1960s and 1970s.[1] These engineers knew theory well enough that they could predict the results of computer simulations. Aerospace industry careers from about 1930 to 1970 included many development projects providing extensive hands-on experience with design, hardware production, software development, and flight testing. This extensive experience enabled engineers to internalize the trade study process. The availability of desk-top computers with powerful mathematical, engineering, and simulation software packages can greatly increase the productivity of engineers. However, there is a danger that current and future engineers may rely too heavily on computer results and not take the time to understand the underlying theory and physics of the problem so that the accuracy and validity of computer results can be confirmed. Formal technical documentation detailing underlying theory and interpretation of results can provide engineers with educational experiences even if senior technical personnel are not available to mentor them.

Engineers should be intellectually curious, constantly asking questions and motivated to spend time doing the work to get the answer and understand it, and understand what factors can change the answer. Good engineers are not interested in simply getting an answer to check the item off on the to-do list and meet a deadline. They perform and contribute meaningful analysis, and do not limit their contributions to mere meeting attendance. Subject matter experts who posses these qualities along with good communication skills can perform the most effective knowledge capture.

---

[1] Computer resources were so scarce in the 1960s that many engineers had to get management permission to use computer time.

Approved for public release via STI DAA 24818. See statement on title page.

Documenting Lessons Learned and Experiences

Documentation is not just a record for contemporaries (other engineers and management at the time of publication), but serves an educational purpose for future engineers.[1] A lot can be learned from failures and development projects that encountered challenges.[2] Most lessons learned that are documented and shared are from bad things that happened in a flight program. However, much can learned from positive experiences as well. It is important not to overlook factors for success and lessons from projects that successfully mitigated risk and met goals. Identifying lessons from successful projects can be more challenging. Personnel tend to spend more time analyzing projects with lots of problems than projects that successfully overcame challenges without negatively impacting mission success, schedule, and budget.

Many that are concerned with knowledge capture and management focus on just-in-time organizational learning through verbal communication (i.e. social networking). This type of effective communication is crucial on a daily basis to ensure the success of spacecraft development and flight operations. However, the effectiveness of social networking is limited by the capacity and accuracy of human memory. Much knowledge and experiences of interest to spacecraft developers and operators is technically detailed and years or decades old. Formal reports and informal memos written in complete sentences with tables, illustrations, and references can preserve technical detail accurately and place it in historical context for future generations of engineers and managers. Well written status reports are a convenient way of recording technical history and preserving the historical context of presentations and reports. Charts with bullet points and spreadsheets omit much background information that may be understood by the original audience, but will not be known to future researchers. This makes charts and spreadsheets difficult to learn from. The knowledge capture approach taken by the authors of the reports detailed in this document preferred formal documents written in paragraphs and ruled out the use of charts with bullet points and spreadsheets.[3]

Program management should provide flexible guidelines for knowledge capture, as opposed to rigid requirements. Guidelines enable subject matter experts and their immediate management to tailor knowledge capture efforts to a specific audience while taking into consideration available skills, time, and budget. Subject matter experts that possess verbal, written, graphic, and inquiry communications skills should be permitted

---

[1] For an interesting overview of spacecraft development lessons from the 1960s, many of which still apply today, see R. D. DeLauer, and E. O. Marriott, "Lessons of the 1960's in Spacecraft Development," *AIAA Space Program Issues of the 70's Meeting*, Seattle, WA, August 28-30, 1967.

[2] Dennehy, Cornelius J., Steve Labbe, and Kenneth L. Lebsock, "The Value of Identifying and Recovering Lost GN&C Lessons Learned: Aeronautical, Spacecraft, and Launch Vehicle Examples," *AIAA Guidance, Navigation, and Control Conference*, Toronto, Ontario, Canada, August 2-5, 2010.

[3] For a discussion of the problems with bullet chart communication see Tufte, Edward R., *The Cognitive Style of PowerPoint: Pitching Out Corrupts Within*, Second Edition, Graphics Press, Cheshire, CT, 2006. Dr. Tufte's books *Envisioning Information* (1990), *Visual Explanations* (1997), *The Visual Display of Quantitative Information* (2001), and *Beautiful Evidence* (2006) (all published by Graphics Press) are excellent sources of information on visual communication, data presentation and analysis. See also page 191 of the Columbia Accident Investigation Board report, "Engineering by Viewgraphs."

Approved for public release via STI DAA 24818. See statement on title page.

to apply those skills to knowledge capture projects. On the other hand, subject matter exports that do not possess these skills should not be expected to perform knowledge capture at a high level. Writing extensive, book length works or even formal reports may not be possible in some cases. Execution of knowledge capture at the subject matter expert level should not be contingent on completion of lengthy development and implementation of high level knowledge management policies, processes, and acquisition of information technology (computer hardware and software).

It is important to document experiences as well as lessons learned. Experiences can be just as informative and are important for transferring knowledge to less experienced engineers and managers. Experiences contain insight that cannot be found in a textbook or a list of lessons learned. Furthermore, it is easier to tell a story and capture the attention of a reader or listener with an experience than with discrete lessons learned.[4] People also tend to remember stories better than isolated statements and facts. Lessons and experiences should be clearly and objectively communicated so that they will be easy for future personnel to understand. Some attempts to communicate lessons are too technical and abstract to be understood by someone that was not a participant. Providing technical and historical background and context is an important part of effectively telling a story.[5]

Effective knowledge capture does not rely solely on the memories of people. Primary sources such as formal reports, memos, presentations, status reports, and meeting minutes are valuable sources of information. Interviewing skills are necessary for drawing key insights and lessons (tacit knowledge) out of subject matter experts that do not possess knowledge capture skills. When performing knowledge capture some subject matter experts often focus on commonly known history, systems descriptions, or process descriptions. They may not recognize key insights (tacit knowledge) that are important for a less experienced person to gain understanding. What is obvious to a senior subject matter expert may be a revelation to a person with less experience.

Below are four steps for the process of capturing and managing lessons learned and experiences. Some organizations focus most of their efforts on the capture and sharing/retrieval steps. However, adequate attention must be paid to the identification and creation steps to ensure that quality knowledge is captured that will be informative, educational, and useful. Ideally, authors should be technical subject matter experts or managers who have the communication and reasoning skills to perform the identification and creation steps. Advice on how to execute the below steps can be sought from people

---

[4] For an excellent example of an experience report with lessons learned see: Nufer, B., *Hypergolic Propellants: The Handling Hazards and Lessons Learned From Use*, Joint JANNAF Interagency Propulsion Committee, 25th Safety and Environmental Protection Joint Subcommittee Meeting, December 6-10, 2010, Orlando, FL. Available from the NASA Technical Reports Server on the internet at the time of this publication.

[5] The *System Failure Case Studies* published by the NASA Safety Center are excellent four page summaries of mishaps. Including in each is background information, a description of the incident, photos and illustrations, proximate causes, underlying issues, and applications to future NASA missions. At the time of this publication the case studies resided on the NASA Process Based Mission Assurance website on the internet.

Approved for public release via STI DAA 24818. See statement on title page.

with previous knowledge capture and writing (such as technical writers and editors) experience when defining tasks and negotiating knowledge capture requirements.

1. Identification – Conduct research to identify lessons. This involves primary source materials and interviews. Research and interviewing skills are required to identify key knowledge, experiences, and lessons.

2. Creation – Create a story or narrative using the results of the research in step 1. This requires skills in reasoning and in verbal, written, and visual communication.

3. Capture – Document the story in some form of media (formal report, informal memo, presentation, training material, procedure, case study discussion outline, video, audio, etc.).

4. Sharing/Retrieval – Some examples are discussion forums (Pause and Learn (PaL), peer assist, case studies, etc.), or retrieving some form of media from a library, website, or electronic database.[6]

The success of knowledge capture and management efforts depends on where most of the work is focused, and by keeping the efforts simple. Knowledge management policy creation, process implementation, and information technology procurement is easier if an organization has already demonstrated some success at knowledge capture. When starting knowledge capture efforts use existing forums for verbal sharing and software tools for electronic capture and electronic sharing and retrieval steps, rather than delaying efforts until new process and policies are in place and budget is allocated for procurement. Much knowledge capture and management can be accomplished by working smarter with resources that are already available.

---

[6] For information on the NASA Pause and Learn (PaL) process see: Rogers, Edward W., "Knowledge Management: Case Study of Building a Learning Organization: Goddard Space Flight Center," in *System Health Management: with Aerospace Applications*, edited by Stephen B. Johnson et al, John Wiley & Sons, Ltd, Chichester, United Kingdom, 2011, pages 65-75.

Approved for public release via STI DAA 24818. See statement on title page.

Choosing Documentation to Preserve

Much knowledge capture involves identifying key documentation and preserving it in a fashion that will facilitate preservation and retrieval by future personnel that may not be familiar with what was saved. In a cost and schedule constrained environment such an approach may be preferable to capturing and documenting tacit knowledge, particularly if the subject matter experts do not possess the communication skills to create knowledge capture documents.

Many basic questions about previous spacecraft and missions can be answered by simple documents such as press kits, mission ground-rules and constraints documents, and crew timelines. Presentations are often not as valuable as formal reports since they do not place the topic in the overall context of the program. The reader is often left with unanswered questions after reading a presentation. These are:

1. Why was this presentation created?

2. What discussion was conducted during the presentation?

3. What action was taken, if any, as a result of the presentation?

The same is true of memos and reports. The presentations, meeting minutes, status reports, memos, and formal reports that are worth saving contain answers to some or all of these questions. Documentation selected for preservation should tell a story.

Rather than approaching document preservation from the perspective of a database gate keeper applying strict requirements, the goal should be to facilitate educational experiences for future engineers, within resource limitations (budget, schedule, computer system capacity, etc.). The fundamental objective of researching legacy programs is to answer the question, "What from the past can be applied to the current project to solve a problem or mitigate risk?" Research is also conducted to help those less familiar with space vehicle design and spaceflight operations understand the complexity and challenges involved. The below questions can help pinpoint what to save for future researchers.

1. What did we do?

2. Why did we do it?

3. When did we do it?

4. How did we do it?

5. Why did we do it that way?

6. What happened?

7. What challenges did we encounter?

8. What did we learn?

9. Is there something we wish we would have done differently?

The above questions can help identify what memos, presentations, and documents are worth saving. However, short memos (one to three pages) may need to be created that tie the documents together and help future engineers see the big picture, or how the documents all fit together to tell a story. People who have worked in a program for many years may understand the connections when looking at documentation. However, these connections may not be obvious to future personnel.

It is impossible to predict what documentation will be important to someone 5 or 30 years from now. One cannot anticipate what the future programs will be, what their requirements will be, or what problems future personnel will be trying to solve. Something that is trivial or obvious to someone working in a flight program may not be understood or recognized by someone in the future. Things that are easily understood by current personnel may be confusing to future personnel. It is impossible to predict what the questions or points of confusion will be when future personnel attempt to solve technical problems and research legacy documentation.

Obviously, not all documentation generated by a flight program can be saved.[1] Binders, filing cabinets, and folders on computer hard drives may be loaded with data but with little or no supporting documentation to explain the data and provide context. There may be a cryptic memo or informal notes accompanying the data. Those that generated the data may not even fully recall what purpose the data served or be able to recall key insight that the data provided. Such data may not be worth preserving.

If some documents are already preserved in formal archives (such as a NASA center library) they do not need to be saved in another archive. References to such documents can be included in document overviews that are written to help future researchers understand the importance of the document collection. However, if a document is already saved in a formal archive and it is an important part of the technical story, it could be saved again in a different place to make the job of future researchers easier. Including the important evidence and sources in one archive makes the future research process easier and saves time, and does not require a large amount of time of the personnel performing the initial knowledge capture. Meeting minutes and associated presentations should be saved together. Researchers are often trying to put together a story using primary and secondary sources that are not as extensive or complete as they would like for them to be. How useful archived documentation is to future personnel depends on how well it tells the story.

---

[1] An extensive document selection process and archive effort was created and executed by USA Flight Design and Dynamics personnel at the end of the Shuttle Program. See Lufkin, Douglas (Barrios Technology), "Shuttle Knowledge Capture, Flight Design & Dynamics," presentation to the Shuttle Knowledge Capture Working Group, March 11, 2010.
Approved for public release via STI DAA 24818. See statement on title page.

Near the end of the Shuttle Program the United Space Alliance Flight Design and Dynamics department created 17 compilation JSC documents to preserve selected primary source material in an organized fashion. A compilation volume consists of a large number of memos, reports, and presentations on a particular topic. Twelve of the volumes concern rendezvous, one navigation, two Draper Laboratory papers, and the final two concerned Space Shuttle Powered Explicit Guidance (PEG). A document overview chapter written by the editor provides the reader with a road map to understanding the content and significance the memos, reports, and presentations. Well written and informative introductions and document overviews are the key to making document compilations useful and understandable to future engineers. This road map explains the significance of the materials and places them in historical context in terms of vehicle development and missions that were flown. Placing the volumes in the NASA/JSC Scientific and Technical Information Center (STIC) under a JSC document number preserved key memos, reports, and presentations that might otherwise end up in the recycling bin. While these materials could have been archived, they would have been less accessible to future engineers performing research. The compilation volumes were scanned to Portable Document Format (PDF), bookmarked, and distributed on DVD to provide on-demand mouse click access that facilitates learning experiences for engineers. This is a low cost method of preserving key primary source materials for future generations of engineers. The volumes also include lists of references to other related documents located in the NASA/JSC STIC.

Lessons Learned and Best Practices Reports

This chapter concerns reports whose primary content was lessons learned and the stories associated with them. Three of the four reports pertain to knowledge management and educating engineering and management personnel on lessons learned to promote development of a good flight safety culture.

Lessons Learned From Seven Space Shuttle Missions

- Goodman, John L., *Lessons Learned From Seven Space Shuttle Missions*, NASA Contractor Report NASA/CR-2007-213697, NASA Johnson Space Center, January 2007. 52 pages.

During the 1992 to 1998 time period the author participated in the investigation of five incidents of anomalous shuttle flight software performance. None of the incidents presented a threat to safety of flight. The author had accumulated and preserved a large amount of primary source material and was motivated to write the report to pass on lessons that could reduce risk in spaceflight programs. The source material included memos and presentations representing the technical expertise and observations of many team members. The author also interviewed other personnel that were involved.

The 52 page NASA contractor report details the causes, mission impacts, and lessons learned from the five incident investigations that the author participated in plus two other investigations that the author was not involved in. Descriptions of each of the seven missions were included. Factors that led to the successful mitigation of the problems were also highlighted. While some technical details had to be omitted to avoid export control concerns, the lessons learned were presented in their entirety.

Table of Contents

- Introduction
- Rendezvous Target Failure On STS-41B
- Rendezvous Trajectory Dispersion On STS-32
- Rendezvous Lambert Targeting Anomaly On STS-49[1]
- Rendezvous Lambert Targeting Anomaly Before STS-51[1]
- Zero Doppler Steering Maneuver Anomaly Before STS-59
- Excessive Propellant Consumption During Rendezvous On STS-69[2]
- Global Positioning System Receiver and Associated Shuttle Flight Software Anomalies on STS-91[3]
- Overall Lessons Learned
- Conclusion

---

[1] For primary source material see Goodman, John L., (editor), *STS-49 Lambert Targeting Anomaly and Aftermath*, JSC-49710, Flight Design and Dynamics Division, NASA JSC, May 2003.

[2] For primary source material see:
  Goodman, John L., (editor), *Space Shuttle Lambert Cyclic Guidance*, JSC-49709, Flight Design and Dynamics Division, NASA JSC, May 2003.
  Meissen, Thomas J., *Space Shuttle Lambert Guidance Improvement, SCR 92843/93009 OI-32*, JSC-49830, NASA/JSC Flight Design and Dynamics Division, October 28, 2006.

[3] For primary source material see Goodman, John L., (editor), *Space Shuttle GPS Lessons Learned*, JSC-49712, NASA JSC Mission Operations Directorate, June 2003.

Approved for public release via STI DAA 24818. See statement on title page.

Flight Safety Education for Current and Future Space Flight Programs

- Goodman, John L., and David P. Dannemiller, "Flight Safety Education For Current and Future Space Flight Programs," *Proceedings of the AIAA Space 2006 Conference*, AIAA, Reston, VA, 2006. 21 pages.

This paper reviews the rationale behind and methodology of the NASA/JSC Mission Operations Directorate (MOD) Flight Safety Seminars. The Flight Safety Awareness Seminar has been held as a part of the yearly NASA Johnson Space Center Safety Day activities since October of 1998. The seminar was motivated by the STS-91 (June 1998) GPS incident.[1] The seminar's primary purpose is to increase flight safety awareness through discussion of accidents, their causes, and lessons learned. In addition, the seminar is designed to motivate and empower personnel to identify potential risks and raise issues in a constructive manner so that they can be investigated before a mishap occurs. Particular attention is paid to what forms of intervention could have prevented the accident and what lessons can be applied to the NASA/JSC mission operations environment so future mishaps can be prevented.[2]

The paper provides details on several incidents examined in the seminars. These include the September 2004 *Genesis* spacecraft crash, loss of the Space Shuttle *Columbia*, the June 1996 Ariane 5 Flight 501 launch failure, the August 1997 loss of the *Lewis* spacecraft, the April 1994 Blackhawk helicopter friendly fire incident over the northern Iraqi no-fly zone, the January 1995 X-31 crash at Edwards Air Force Base, the March 2001 crash of a Gulfstream III in Aspen, Colorado, and the November 1999 Texas A&M bonfire collapse.

Topics Covered

- Introduction
- Shuttle Upgrade Challenges Led to Establishment of the Safety Seminars
- Seminar Source Material
- Seminar Methodology
- Seminar Evolution, Lessons Learned and Best Practices
- Expansion of the Flight Safety Education Effort Since 1998
- Conclusions
- Appendix A – Nine Incidents Covered in the Seminars
- Appendix B – Incident Investigation Report Completion and Publication Dates

---

[1] Goodman, John L., *Lessons Learned From Seven Space Shuttle Missions*, NASA Contractor Report NASA/CR-2007-213697, NASA Johnson Space Center, January 2007.

[2] The Columbia Accident Investigation Board stated on page 183 of their report that "NASA emphasizes safety as well, but training programs are not robust and methods of learning from past failures are informal." The MOD Flight Safety Seminars, begun over four years before the loss of *Columbia* and her crew, are a formal method of training personnel to recognize potential flight safety risks and promote a healthy flight safety culture.

Approved for public release via STI DAA 24818. See statement on title page.

Best Practices for Researching and Documenting Lessons Learned

- Goodman, John L., *Best Practices for Researching and Documenting Lessons Learned*, NASA Contractor Report NASA/CR-2008-214777, NASA Johnson Space Center, March 2008. 20 pages.

When USA Flight Design began to perform navigation and rendezvous lessons learned research for the NASA/JSC Flight Design and Dynamics Division, no documents could be found providing guidelines or best practices for conducting such an activity. As a result, lessons learned, factors for success, and observations were collected from all participants (NASA and contractor, management and technical) after publication of several reports. These interviews were used to define best practices for researching and documenting lessons learned and experiences. The purpose of this report is to assist government and contractor organizations to perform knowledge management of lessons learned and experiences.

<u>Table of Contents</u>

- Introduction
- Defining Report Requirements, Project Organization, and Schedule
- Collection and Analysis of Source Material
- Writing and Integrating the Report
- Review and Revision of the Report
- Conclusion
- References

Knowledge Capture and Management for Space Flight Systems

- Goodman, John L., *Knowledge Capture and Management for Space Flight Systems*, NASA Contractor Report NASA/CR-2005-213692, NASA Johnson Space Center, October 2005. 24 pages.

This report stressed the importance of knowledge capture and provided examples of low cost knowledge capture and management activities performed at the NASA Johnson and Kennedy Space Centers.

Topics Covered

- Introduction
- Knowledge Capture and Management Is Important
- Why Knowledge May Not Be Captured or Accessible
- Legacy Vehicles and Systems
- Managing Talent and Changing Culture
- Improving Knowledge Capture and Management In Future Programs
- Conclusion

Navigation and GPS Reports

The Space Shuttle GPS integration, flight test, and certification was a successful project that resulted in improved and excellent on-board navigation performance. The project also provided many lessons learned due to the technical and programmatic challenges that had to be overcome. Six of the reports covered in this chapter concern the GPS project. Two cover shuttle navigation software upgrades not associated with GPS. *Navigation Technical History With Lessons Learned* also covers non-GPS on-board and ground navigation.

Space Shuttle GPS Operating Characteristics

- Goodman, John L., and Mark C. Morris, *Space Shuttle GPS Operating Characteristics, 19th Edition*, JSC-35082, NASA/JSC Flight Dynamics Division, July 2011. 194 pages.

GPS Operating Characteristics contained information on GPS receiver functionality, performance, definitions of commands and data presented to the crew and Mission Control personnel. It also contained shuttle software details, and information on other GPS receivers and GPS/Inertial Navigation System (INS) (or SIGI) units under consideration by NASA/JSC for use in space. This included the International Space Station and X-38 GPS receivers. Information included in the document was harvested from interaction with the GPS receiver vendor and other shuttle GPS personnel, post flight analysis, meeting discussions, emails, and presentations.

Background on This Report

Several weeks before the first flight of the Miniaturized Airborne GPS Receiver/Shuttle (MAGR/S) GPS receiver on STS-79 (September 1996) Ascent/Entry Guidance and Procedures Officer (GPO) Kelly Beck asked John Goodman to provide information on GPS receiver data and other parameters. Although the MAGR/S was not yet certified for operational use there were contingency procedures for it to be used with the Backup Flight System (BFS) in a worst case navigation scenario. In response to the request a four page memo titled *OI-25 GPS Display Definitions* was distributed to shuttle GPS and Mission Control Center personnel on September 3, 1996.

Due to the difficulty in obtaining formal documentation on MAGR/S functionality the memo quickly became a living knowledge capture document and was eventually renamed *GPS Operating Characteristics*. By the time of the publication of the 18th edition on May 24, 2001, it had grown from 4 pages to 174 pages. Each edition of the document was distributed to Shuttle Program personnel concerned with GPS and shuttle navigation in general.

While *GPS Operating Characteristics* was widely used by Shuttle Program personnel, it was never an official document. In support of end-of program knowledge capture Mark Morris of the United Space Alliance Flight Design and Dynamics Department prepared an updated 19th edition for publication and preservation as a JSC document in July of 2011.

Space Shuttle GPS Compilation

- Goodman, John L., (editor), *Space Shuttle GPS Lessons Learned*, JSC-49712, NASA JSC Mission Operations Directorate, June 2003. 206 pages.

Most of the compilation consists of presentations concerning the June 1998 STS-91 GPS incident.[1] These presentations concern the causes of the incident and corrective actions that were taken. They were presented to senior Shuttle Program management. It also contains shuttle GPS lessons learned papers and presentations. The lessons learned were later included in the NASA contractor report *GPS Lessons Learned From the ISS, Space Shuttle, and X-38*.[2]

---

[1] Goodman, John L., *Lessons Learned From Seven Space Shuttle Missions*, NASA Contractor Report NASA/CR-2007-213697, NASA Johnson Space Center, January 2007.

[2] Goodman, John L., *GPS Lessons Learned From The ISS, Space Shuttle and X-38*, NASA Contractor Report NASA/CR-2005-213693, NASA Johnson Space Center, November 2005.

Approved for public release via STI DAA 24818. See statement on title page.

Improvement of Space Shuttle Time to Node Computation

- Goodman, John L., *Improvement of Space Shuttle Time to Node Computation*, JSC-49766, NASA JSC Mission Operations Directorate, July 2003. 56 pages.

The Space Shuttle Relative Navigation (REL NAV) crew display (also called SPEC 33) contained a countdown timer, called time-to-node, to the next orbiter crossing of the target spacecraft orbital plane. This could be used by the crew after the Mid-course Correction-1 (MC-1) burn as cue for zeroing out-of-plane velocity by manually commanding Reaction Control System (RCS) jet firings.

The original (late 1970s) computation for time-to-node differenced current and previous values of out-of-plane position, divided by the time difference, to determine out-of-plane velocity. State vector updates from Kalman filtering of rendezvous radar data resulted in an inconsistency in the position values used to compute out-of-plane velocity, rendering the time-to-node parameter unusable by the crew.

A Flight Software Change Request (CR) was approved for flight software Operational Increment-30 (OI-30) on September 6, 2001, to improve the accuracy of the time-to-node computation. The improvement involved replacing the numerical computation of out-of-plane velocity with an exact analytical solution. This document detailed the derivation of the new out-of-plane velocity equation, and the changes made to the shuttle flight software. The first flight of OI-30 was the STS-114 flight of *Discovery* (July-August 2005).

Space Shuttle Rotation, Nutation and Procession (RNP) Matrix Computation

- Brownd, Joseph E., *Space Shuttle Rotation, Nutation and Procession (RNP) Matrix Computation (CR 92329E)*, JSC-49834, September 2003. 54 pages.

This report documented the theory behind an upgrade to the Space Shuttle on-board flight software Rotation, Nutation and Procession (RNP) matrix algorithm. The new on-board RNP computation method eliminated the need to compute up to ten RNP matrices and associated time epochs per year, and eliminated the need to patch the flight software in the event of concerns with RNP staleness and associated navigation errors. The first flight of the new algorithm, in software version OI-29, was on *Atlantis* (STS-110) in April of 2002.

GPS Lessons Learned From the ISS, Space Shuttle, and X-38

- Goodman, John L., *GPS Lessons Learned From The ISS, Space Shuttle and X-38*, NASA Contractor Report NASA/CR-2005-213693, NASA Johnson Space Center, November 2005. 120 pages.

This report is a compilation of most material written by the author on Global Positioning System (GPS) navigation from January 2001 to November 2005. It details the GPS navigation integrations on the Space Shuttle, ISS, and X-38. Lessons learned from these three programs are presented.

Table of Contents

- Preface
- Introduction
- The Software Nature of Satellite Navigation
- GPS Lessons Learned
- International Space Station
- The Space Shuttle and GPS
- Other GPS Receivers On The Space Shuttle
- Shuttle Space Integrated GPS/INS
- X-38/Crew Return Vehicle
- Rationale Behind a Notional Shuttle GPS Receiver Upgrade
- Summary

Background on This Report

In the early 1990s the Shuttle Program began to re-investigate the possibility of replacing the onboard Tactical Air Navigation (TACAN) units used during entry with GPS receivers.[1] The Program favored the use of an existing GPS receiver designed for, and flight-proven by, military aviation. Such a receiver had the benefits of authorized operation, ability to accept inertial aiding, resistance to jamming, and an existing logistics support base. At the time there were no space GPS receivers in production that were designed to support the entire Shuttle flight envelope (powered flight during ascent, on-orbit, atmospheric re-entry) with these design features, nor were any of the existing space receivers compatible with the Shuttle from a form-and-fit perspective. Use of avionics and other devices from atmospheric flight applications was not new to the Shuttle Program. Many systems on the Shuttle were adapted from other atmospheric flight vehicles in the 1970s. However, GPS receivers contained far more software than previous avionics devices that had been adapted for use on the Shuttle. The Shuttle GPS upgrade project encountered more technical and project management challenges than anticipated,

---

[1] Studies of using GPS navigation on-board the Space Shuttle were also conducted in the mid and late 1970s, and during the 1980s.

Approved for public release via STI DAA 24818. See statement on title page.

despite the rapidly increasing and successful use of GPS technology worldwide in numerous applications.

By 1996, concerns about the software intensive nature of the Shuttle GPS receiver led concerned personnel (both engineering and management) to seek experiences and lessons learned from other flight programs. There was a considerable amount of material available advocating the use of off-the-shelf software and hardware, but at the time (mid 1990s) little guidance was available on selection, integration, testing, and certification of software intensive off-the-shelf or modified-off-the-shelf units. Papers concerning the use of off-the-shelf hardware and software were very abstract, difficult to understand, and provided little practical guidance on selection and implementation. GPS project personnel found it difficult to locate relevant lessons learned from projects that encountered technical and programmatic challenges.

Formal reports from accident investigations and spacecraft failures were available (Ariane 501, Mars Climate Orbiter, Mars Polar Lander, Lewis, WIRE, etc.) and were reviewed by some GPS project personnel. Some of these reports highlighted challenges that were also of concern to Shuttle GPS project members. These challenges included software and spacecraft development using software and hardware originally developed for other flight programs. However, project personnel were not able to effectively communicate that parallels existed between the causes identified in the accident reports and problems observed in the Shuttle GPS project.

In June of 1998, during the flight of STS-91, interaction between a malfunctioning GPS receiver and two previously unknown software anomalies in the Shuttle flight computer resulted in a loss of communication with *Discovery* for approximately 30 minutes.[2] The STS-91 incident and the difficulty in locating useful and applicable lessons learned from other projects motivated personnel to formally document lessons learned from the Shuttle GPS project.

Many project GPS project personnel had been identifying technical and programmatic issues and extracting lessons learned from them for several years, but until the STS-91 loss of communications incident no lessons learned had been formally documented. Soon after STS-91 several program internal lessons learned presentations were created. The collection of observations and lessons learned was conducted so as to protect the identity of the contributors. Some of the internal presentations were not widely distributed due to the sensitive nature of the lessons and observations. A NASA reorganization of the Shuttle GPS project made use of these observations and lessons. Shuttle GPS was certified for operational use in 2002, after a three-year delay.

Later, further collection of GPS lessons learned was performed by United Space Alliance so that they could be disseminated to a wider NASA and aerospace industry audience. Lessons were collected from the STS-91 investigation presentations, discussions at

---

[2] Goodman, John L., *Lessons Learned From Seven Space Shuttle Missions*, NASA Contractor Report NASA/CR-2007-213697, NASA Johnson Space Center, January 2007.

Approved for public release via STI DAA 24818. See statement on title page.

meetings, and interviews with managers and engineers on an individual basis. Lessons from the ISS and X-38 GPS projects were later collected in the same manner.

The GPS project consisted of a small team of NASA and contractor personnel. Many project members had participated in the project for most of its existence. Due to the organizational and technical challenges that were encountered, lessons learned were a frequent topic of discussion long before an attempt was made to document them. Memories were fresh, so identifying and harvesting lessons learned was straightforward, either through interviews or by simply listening to discussions during meetings. However, not all technical personnel fully understood the programmatic decisions made by Shuttle Program management early in the GPS project (before 1996). This factor had to be taken into account when evaluating their observations.

Seven conference papers and one magazine article were published, highlighting various aspects of the Shuttle, ISS, and X-38 GPS projects, in addition to lessons learned.[3] Once drafts of papers were completed, they were provided to both Mission Operations and Engineering Directorate personnel (NASA and contractor, management and technical) for review. This ensured that all aspects of the project were accurately documented and that the papers represented the views of JSC personnel and their supporting contractors in various organizations. Since the papers were written for external publication, care was taken to write the lessons to allay export control concerns. Subject matter experts provided quick, insightful, and accurate reviews of draft papers. Some material from the first paper was placed in the NASA Lessons Learned System soon after publication to ensure availability of the lessons to a wider audience.[4]

---

[3] Goodman, John L., "Parallel Processing: GPS Augments TACAN in the Space Shuttle," *GPS World*, Volume 13, Number 10, October 2002. See also http://www.gpsworld.com/government/aviation-space/parallel-processing-767, accessed July 20, 2011.

Goodman, John L., "Space Shuttle Navigation in the GPS Era," *Proceedings of the National Technical Meeting 2001*, Institute Of Navigation, Long Beach, CA, January 22-24, 2001, pages 709-724

Goodman, John L., "Lessons Learned From Flights of "Off the Shelf" Aviation Navigation Units on the Space Shuttle," *Joint Navigation Conference*, Orlando, Florida, May 6-9, 2002.

Goodman, John L., "GPS In Earth Orbit – Experiences From The Space Shuttle, International Space Station And Crew Return Vehicle Programs," *Proceedings of the 2002 Core Technologies for Space Systems Conference*, Colorado Springs, CO, November 19-21, 2002.

Goodman, John L., "The Space Shuttle and GPS – A Safety-Critical Navigation Upgrade," *COTS-Based Software Systems*, edited by H. Erdogmus and T. Weng, Springer-Verlag Lecture Notes in Computer Science Volume 2580, Springer-Verlag, Berlin, Germany, 2003, pp. 92-100.

Goodman, John L., "A Software Perspective On GNSS Receiver Integration and Operation," *Satellite Navigation Systems: Policy, Commercial and Technical Interaction*, edited by M. Rycroft, Kluwer Academic Publishers, Dordrecht, The Netherlands, 2003, pp. 119-126.

Goodman, John L., "A GPS Receiver Upgrade for the Space Shuttle - Rationale and Considerations," Paper AIAA-2004-3911, *40th AIAA/ASME/SAE/ASEE Joint Propulsion Conference and Exhibit*, Fort Lauderdale, FL, July 11-14, 2004.

Goodman, John L., "Applications of GPS Navigation to Space Flight," *2005 IEEE Aerospace Conference*, IEEE, New York, NY, 2005.

[4] Goodman, John L. *Lessons Learned From Flights of "Off the Shelf" Aviation Navigation Units on the Space Shuttle*, NASA Public Lessons Learned Entry 1370, NASA Lessons Learned Database, NASA Engineering Network, June 11, 2002. http://www.nasa.gov/offices/oce/llis/1370.html. Accessed July 20, 2011.

Approved for public release via STI DAA 24818. See statement on title page.

The externally published papers were later integrated and re-formatted into a 120 page NASA Contractor report *GPS Lessons Learned From The ISS, Space Shuttle and X-38*. The report was made publicly available through the NASA Technical Reports Server on the internet. In addition to the publicly available papers, many internal presentations concerning project lessons learned, the STS-91 GPS incident, and the subsequent investigation were compiled into a JSC document and placed in the JSC technical library to ensure access and preservation of original source material.[5]

The author later encouraged the NASA/JSC Engineering Directorate to publish a NASA Technical Publication documenting lessons learned from the ISS GPS project.[6] This report provides more of a procurement and development perspective on ISS GPS challenges encountered by NASA/JSC.

---

[5] Goodman, John L (editor), *Space Shuttle GPS Lessons Learned*, JSC-49712, Flight Design and Dynamics Division, Mission Operations Directorate, NASA Johnson Space Center, June 2003. Available from the NASA/JSC Scientific and Technical Information Center.

[6] Gomez, Susan, *Three Years of Global Positioning System Experience on International Space Station*, NASA Technical Publication NASA/TP-2006-213168, NASA Johnson Space Center, August 2006. See the NASA Technical Reports server at http://ntrs.nasa.gov/, or the Johnson Technical Reports server at http://ston.jsc.nasa.gov/collections/TRS/. Accessed July 20, 2011.

Navigation Technical History with Lessons Learned

- Goodman, John L., Toni Deboeck, Daniel Dyer, Robert Goyen, Michael Nishizaki, Crystal Ramirez, Devon Sanders, Sherrie Santos, Matthew Shaver, and Samuel Welsh, *Navigation Technical History with Lessons Learned*, JSC-63653, Flight Design and Dynamics Division, Mission Operations Directorate, NASA Johnson Space Center, April 2007. 382 pages.

This report covers human spacecraft navigation techniques, historical navigation architectures, lessons learned, flight performance history, evolution of navigation software, and legacy best practices from previous spaceflight programs. It was researched and written for Constellation Program risk mitigation.

Table of Contents

- Navigation Lessons Learned and Observations
- Mercury Navigation
- Gemini Navigation
- Apollo Navigation
- Shuttle Navigation
- Shuttle Ground Navigation Software Development
- International Space Station Navigation
- Concluding Thoughts
- References

Background on This Report

Over an eight-month period a team of Mission Operations navigation specialists located, obtained, and studied hundreds of documents from the Mercury, Gemini, Apollo, Space Shuttle, and ISS Programs. Source material was abundant and dated back to the late 1950s. The team also took advantage of a large body of primary source material that had been collected by a previous project. This saved a considerable amount of time. These documents included internal NASA reports, memos, presentations, and published technical papers and books. However, there was no central repository for historical technical documentation; therefore, identifying, locating, and obtaining certain historical documents was a tedious and time consuming process. Source material was listed in a references section by flight program. Available source documents were placed on compact disks and delivered to the NASA customer along with the final report. This will permit future researchers to access original source documents in a timely manner.

On-board and ground system anomalies were included to enable personnel, including those new to the human flight program, to identify long-term trends of technical anomalies and organizational challenges. The document was designed to provide the Constellation Program (and future development and flight programs) ready access to such information, as much of the source material on this topic is not centrally located or organized. The document also enables new personnel, who may never have had the opportunity to work with veteran engineers that developed on-board and ground navigation software, architecture, and techniques in the 1960s and 1970s, to gain insight into accomplishments, milestones, lessons learned, and best practices from that era. The 382 page report was used as a reference by NASA and contractor personnel working in the Orion and Constellation programs.

Operational Aspects of Space Shuttle GPS

- Goodman, John L., "Space Shuttle Navigation in the GPS Era," *Proceedings of the National Technical Meeting 2001*, Institute Of Navigation, Long Beach, CA, January 22-24, 2001. 17 pages.

- Goodman, John L., and Carolyn A. Propst, "Operational Use of GPS Navigation for Space Shuttle Entry," *IEEE/ION PLANS 2008 Conference*, Monterey, CA, May 5-8, 2008. 13 pages.

Space Shuttle Navigation in the GPS Era describes how Space Shuttle on-board and ground navigation was and was not changed by the addition of GPS receivers to the Space Shuttle. Pre-GPS and post-GPS shuttle navigation is described by flight phase (ascent and post insertion, orbit coast, rendezvous, deorbit, and entry). The integration architecture, state replacement, is described along with the rationale for its selection. Also covered are integration architectures that were not chosen, receiver modification for the Space Shuttle, flight test results, flight tests of the Space Integrated GPS/INS (SIGI) unit, and lessons learned.

Operational Use of GPS Navigation for Space Shuttle Entry describes the operational use of one GPS receiver and three TACAN units by the Space Shuttles *Atlantis* and *Discovery*, and operational use of three GPS receivers (no TACANs) by *Endeavour*. The single string test flights (called GPS ramp up) by *Atlantis* and *Discovery* leading to the first flight of no TACANs and three GPS receivers by *Endeavour* on STS-118 (August 2007) are described. Plans for nominal and contingency use of single string GPS is covered. The paper also includes a comparison of entry navigation on the Mercury, Gemini, Apollo Command Module, and Space Shuttle vehicles.

Approved for public release via STI DAA 24818. See statement on title page.

Ionopheric Scintillation and GPS

- John L. Goodman and Leonard Kramer, "Scintillation Effects On Space Shuttle GPS Data," *Proceedings of the ION 2001 National Technical Meeting*, Long Beach, CA, January 22-24, 2001, pages 742-752.

- Leonard Kramer and John L. Goodman, "Ionospheric Instability Observed in Low Earth Orbit Using Global Positioning System," *AIAA Journal of Spacecraft and Rockets*, Vol. 42, No. 1, January-February 2005, pages 124-131.

Two articles on the challenges of GPS navigation appeared in *Aviation Week & Space Technology* magazine in December 1997.[1] As a result of these articles the NASA/JSC Mission Operations Directorate held a series of splinter Ascent/Entry Flight Techniques meetings in early 1998 to review the approaching replacement of shuttle TACAN units with GPS receivers. The splinters were chaired by Ascent/Entry Flight Director and future Shuttle Program manager Wayne Hale. Mr. Hale later informed Shuttle Program management that there was no significant reason to delay replacement of TACAN units with GPS receivers.[2] However, there were several recommendations for further investigative work, one of which was a better understanding of ionospheric scintillation at the approaching solar maximum and the potential impact on shuttle GPS navigation performance. During the remainder of 1998 United Space Alliance personnel investigated ionospheric scintillation effects on GPS receiver performance with the assistance of GPS and ionospheric scintillation specialists in industry and academia.

On the evening of November 3, 1998, during the flight of STS-95, NASA Mission Control Ascent/Entry Guidance and Procedures Officer Glenn Pogue noted two periods of noisy GPS velocity. The phenomenon was observed on two consecutive orbits, as the Shuttle *Discovery* was off the west coast of South America, during the early evening hours. The noisy velocity was in the range of 5 to 7 feet/second for about 5 minutes. Based on the scintillation study conducted the previous spring, ionospheric scintillation was identified as the most probable cause. The papers *Scintillation Effects On Space Shuttle GPS Data* and *Ionospheric Instability Observed in Low Earth Orbit Using Global Positioning System* detail the analysis that was performed to confirm that the velocity noise was caused by ionospheric scintillation. The physics underlying ionospheric scintillation is reviewed and why the velocity noise occurrences were judged not to present a risk to safety of flight or mission success once GPS replaced TACAN.[3]

---

[1] Nordwall, Bruce D., "GPS Success Sparks New Concerns for Users," and "Solar Storms Threaten GPS Reception," *Aviation Week & Space Technology*, Vol. 147, No. 22, December 1, 1997, pages 58-60 and pages 61-62.

[2] Hale, N. W., "Operations Review: GPS Replacement for TACANs," Flight Director Office, NASA JSC Mission Operations Directorate, March 13, 1998.

[3] The velocity noise was found to be a function of shuttle velocity. In the unlikely event that ionospheric scintillation occurred during landing the noise would be too low to negatively impact navigation performance. While incorporation of GPS state vectors into the shuttle navigation system while on-orbit was certified the Shuttle Program chose not to develop nominal procedures for on-orbit incorporation to save money, since the Program was scheduled to end in 2010.

Approved for public release via STI DAA 24818. See statement on title page.

Navigation Papers by Gene Brownd and Dr. Leonard Kramer

- Goodman, John L. (editor), *Navigation Papers by Gene Brownd and Leonard Kramer*, JSC-35084, Flight Dynamics Division, Mission Operations Directorate, NASA Johnson Space Center, June 2011. 1178 pages.

This compilation preserves memos and presentations by Gene Brownd and Dr. Leonard Kramer from the years 1994-2005. Mr. Brownd and Dr. Kramer worked in the United Space Alliance Flight Design and Dynamics Department Navigation group. Both were recognized and highly regarded experts on Kalman filtering. Mr. Brownd's papers concern Kalman filtering, Space Shuttle GPS receiver navigation performance, quaternions, GPS antenna modeling, GPS receiver navigation error modeling, and IMU alignment. Dr. Kramer's presentations and memos detail the theoretical development and functionality of the Mission Control Spacecraft Position Optimal Tracking (SPOT) filter. SPOT was used in Mission Control to provide accurate orbit determination of the Space Shuttle and ISS by filtering on-board GPS receiver position vectors.

Rendezvous and Proximity Operations Reports

Rendezvous and proximity operations were an integral part of many shuttle missions from June 1983 through the end of the program in July of 2011. The successful development and demonstration of rendezvous techniques during Project Gemini was essential for the later success of the Apollo lunar missions, as well as the Skylab and Apollo/Soyuz missions. However, shuttle rendezvous and proximity operations presented new technical challenges through the life of the Shuttle Program. These reports provide insight into these challenges and how they were met. The compilation volumes preserve a considerable amount of primary source documentation.

Space Shuttle Rendezvous and Proximity Operations Experience Report

- Goodman, John L., and Kelli S. Wiuff, *Space Shuttle Rendezvous and Proximity Operations Experience Report*, JSC-49626, Flight Design and Dynamics Division, Mission Operations Directorate, NASA Johnson Space Center, February 2003. 192 pages.

Under the Space Launch Initiative (SLI) Program, NASA tasked the USA Flight Design and Dynamics department to conduct an eight-month effort (beginning in June of 2002) to collect experiences, lessons learned, and best practices from 20 years of planning and flying Space Shuttle rendezvous and proximity operations missions.

The report contains an introduction explaining the report, an executive summary, strengths and accomplishments of shuttle rendezvous and proximity operations, an overview of rendezvous and proximity operations phases, and a historical overview of shuttle rendezvous and proximity operations development. Two chapters provide generic descriptions of various aspects of vehicle systems and mission planning using examples from shuttle missions. The examples are detailed in the Supporting Evidence chapter. Tables of in-flight anomalies and shuttle rendezvous mission history were also included. An extensive bibliography cited open literature papers related to shuttle rendezvous and proximity operations. The report was provided to rendezvous personnel working on the Orion Program.

Table of Contents

- Background
- Vehicle Systems/Subsystems
- Mission Planning/Operations
- Supporting Evidence
- Shuttle History Lists
- Further Information

Background on This Report

A significant amount of historical material (post flight reports, mission planning memos, flight techniques presentations, etc.) were collected and examined. This material occupied several shelves of binders and filing cabinet drawers. Most of the material in the report was obtained through examination of the collected documentation.

Some lessons learned were explicitly identified in primary source documents, particularly those dating from 1983 and 1984, when the first shuttle rendezvous and proximity operations missions were flown. Other lessons were extracted and formulated after careful consideration of the material and discussion with Mission Operations subject matter experts. A tracking form was developed for use during the research phase to record lessons learned and observations. The form contained entries for tracking number,

Approved for public release via STI DAA 24818. See statement on title page.

author, date, title, lesson learned, relevant background information, and references. The USA document author and USA project manager had a close relationship with the NASA customer. This permitted frequent discussion of project status and timely resolution of report requirements and scope issues.

The definition of a lesson learned found on the NASA Lessons Learned Information System website in 2002 was used by the project to identify material to be placed in the report:

> *A lesson learned is knowledge or understanding gained by experience. The experience may be positive, as in a successful test or mission, or negative, as in a mishap or failure. Successes are also considered sources of lessons learned. A lesson must be significant in that it has a real or assumed impact on operations; valid in that it is factually and technically correct; and applicable in that it identifies a specific design, process, or decision that reduces or eliminates the potential for failures and mishaps, or reinforces a positive result.*

A best practice was defined as a factor taken into consideration when defining mission plans, flight rules, procedures, and vehicle design. A lesson learned may become a best practice and the original lesson may be forgotten over time. Many subject matter experts felt that documenting only lessons learned would exclude much valuable information. Therefore, the report was titled an "experience report." Historical material was added to place the material in context. Originally, a distinction was to be made in the report between lessons learned, experiences, best practices, and observations (a catch-all category for anything that did not fit in the other three categories). A limited schedule prevented these distinctions from being made in the report.

A 170 page draft report was published for review in early December of 2002. Reviewers included NASA and USA personnel in the Flight Design and Dynamics organization, as well as USA personnel in rendezvous and proximity operations crew training. Of those subject matter experts that participated in interviews and draft reviews, a few had rendezvous and proximity operations experience dating back to the 1960s and 1970s. A slightly larger number were involved in rendezvous and proximity operations before the loss of *Challenger*.

Four formal review meetings were held over a four week period in January and February of 2003. Numerous discussions over email, phone, and in person were also held. Release of the first draft of the experience report prompted reviewers to supply constructive criticism. While meetings to discuss the draft were useful, the most valuable input was obtained from redlined copies or emails containing comments on the drafts. Far more lessons and experiences were harvested in this fashion than from interviews.

Unlike shuttle GPS, rendezvous and proximity operations concerned multiple systems and disciplines, which resulted in more subjects to investigate as compared to the GPS project. Rendezvous and proximity operations subject matter experts found it more difficult to identify lessons learned than GPS project personnel, since they did not have the benefit of a project that faced serious technical and programmatic challenges to motivate them to identify lessons learned. The experience report went through considerable change before publication as a JSC document. Based on discussion during the reviews, the report was reorganized to improve its legibility. Headings and subheadings were redefined. Generic summaries for each subheading were written, with references to supporting evidence in the latter half of the document. The supporting evidence contained specific examples from which the generic statements were derived and cited the sources from which the evidence was obtained.

During the research phase a considerable amount of shuttle rendezvous history material was obtained. Since an easily accessible and readable narrative history of shuttle rendezvous did not exist, a 17 page *AIAA Journal of Spacecraft and Rockets* article and a much longer JSC document was published.[1,2]

---

[1] Goodman, John L., "History of Space Shuttle Rendezvous and Proximity Operations," *AIAA Journal of Spacecraft and Rockets*, Vol. 43, No. 5, Sept.-Oct. 2006, pages 944-959.

[2] Goodman, John L., *History of Space Shuttle Rendezvous, Revision 3*, JSC-63400, Flight Dynamics Division, Mission Operations Directorate, NASA Johnson Space Center, October 2011.

Approved for public release via STI DAA 24818. See statement on title page.

Rendezvous and Proximity Operations Systems Integration

- Goodman, J. L., and J. P. Brazzel, "Rendezvous Integration Complexities of NASA Human Flight Vehicles," *32nd Annual AAS Guidance And Control Conference, Guidance and Control 2009, Advances in the Astronautical Sciences*, Volume 133, Univelt, San Diego, CA, 2009. 17 pages.

This paper was written to counter the perception that rendezvous, proximity operations, and docking is just about relative sensors and propellant optimal trajectories. Topics covered include: 1) rendezvous missions and phases of rendezvous, 2) design, development, test and engineering, 3) chaser and target vehicle integration, 4) vehicle design, 5) integrating rendezvous and proximity operations systems, 6) contingency planning, and 7) integrated teams. The paper also discusses rendezvous systems integration lessons learned from four case studies: 1) late addition of VHF ranging to the Apollo CSM, 2) late recognition of Space Shuttle plume impingement, 3) balancing automation, autonomy, and authority (Soyuz spacecraft), and 4) sensor risk reduction through flight testing.

Space Shuttle Rendezvous Training

- Goodman, John L., *Introduction To Space Shuttle Rendezvous Guidance, Navigation, and Control*, Fourth Edition, JSC-49686, NASA JSC Flight Design and Dynamics Division, November 2009. 420 pages.

This fully illustrated book is a high level introduction to rendezvous trajectories, relative navigation, burn targeting and guidance, and flight control. It was used as a training and reference book by mission planning, Mission Control, and engineering personnel at NASA/JSC. It was also provided to astronauts for supplementary reading.

Table of Contents

- Coordinate Systems
- Launch Windows and Constraints
- Ascent, OMS-2, and the Ground Targeted Phase
- Day of Rendezvous and Separations
- Rendezvous Profile Evolution and Scenarios
- On-Board Rendezvous Targeting and Guidance
- Rendezvous Navigation
- Flight Control During On-board Targeted Phase
- Integrated GN&C During Rendezvous

Background on This Book

Introduction to Space Shuttle Rendezvous Guidance, Navigation, and Control was originally written (mid 1991 to February 1993) to meet rendezvous training needs within the Rockwell Space Operations Company (RSOC) Level 8 guidance, navigation, and control flight software testing group.[1]

Rendezvous training materials available circa 1986-1991 were difficult to use for training entry level engineers to perform verification of shuttle flight software. Many new hires and some Shuttle Program veterans found rendezvous and proximity operations to be a difficult topic to understand. Many aspects of rendezvous profiles, guidance, navigation and control were not covered in existing training books, nor did they tie the components of rendezvous and proximity operations together to give the trainee the big picture.

The text, including style and content, was developed in response to lessons learned over five and a half years of training entry level engineers to perform rendezvous software verification and effectively interface with other elements of the Shuttle Program. Most of the material in the 250 page first edition (published in February of 1993 as a STSOC

---

[1] Level 8 flight software testing was performed on the integrated mass memory for each shuttle mission. The testing was performed in the Software Production Facility, or SPF, at the Johnson Space Center. The SPF contained shuttle General Purpose Computers (GPCs, the AP-101B and later AP-101S).

Approved for public release via STI DAA 24818. See statement on title page.

document, STSOC-TM-001498) was taken from rendezvous training classes held for Level 8 analysts in the spring of 1991.

The book was intended to allow engineering and management personnel to quickly come up to speed through the development of a visual rendezvous vocabulary. Creating pictures required a thorough understanding of rendezvous design and flight software algorithms, necessitating authorship by a subject matter expert. Pre-publication reviewers included Mission Operations Directorate (MOD) Flight Design and Crew Training personnel.

Use of the book quickly spread to other NASA/JSC organizations outside of RSOC Level 8 software verification, making it difficult to determine who was using it. Although the target audience was technical personnel it was provided to Space Shuttle crews for supplementary reading.

In May of 1999 the MOD Rendezvous Guidance and Procedures Office asked the author if an updated, electronic edition could be published. The new edition would reflect rendezvous and docking with the International Space Station and the new rendezvous profile. Work on the new edition began in the fall of 1999. The book was rehosted from MacDraw to Powerpoint, graphics were cleaned up, more realistic crew displays were created, and 117 pages of new material were added. The second edition was published on February 9, 2000, and a third edition with corrections was published on May 17, 2000. It was available online as a pdf file. The third edition was used by some Orion and Constellation personnel as a reference for rendezvous GNC and trajectory concepts.

The 420 page fourth edition (published November 2009) contained corrections that had been identified since the 3rd edition was published in May of 2000. In addition, new material was added that was developed by the author in the 1990s and early part of the first decade of the 21st century. While some of this material was perhaps too detailed for an introduction, it was deemed of value and the author wanted to make it available to future JSC personnel in the interest of knowledge capture. It was published as a JSC document to ensure preservation after the end of the Shuttle Program and availability to personnel across NASA.

Visiting Vehicles Rendezvous Training

- Goodman, John L., *Introduction to Relative Navigation Concepts for Visiting Vehicles Officers, First Edition*, JSC-36583, Flight Dynamics Division, Mission Operations Directorate, NASA Johnson Space Center, September 2010. 182 pages.

- Goodman, John L. *Introduction to Rendezvous Burn Targeting and Guidance Concepts for Visiting Vehicles Officers, First Edition*, JSC-36584, Flight Dynamics Division, Mission Operations Directorate, NASA Johnson Space Center, September 2010. 242 pages.

These books were based on the fourth edition of *Introduction To Space Shuttle Rendezvous Guidance, Navigation, and Control*, but with much new material and significant revision of material from the shuttle book. The books were written for NASA/JSC Mission Operations personnel that were concerned with the rendezvous aspects of Visiting Vehicles (Soyuz, Progress, ATV, HTV, Dragon, Cygnus, etc.).

Tables of Contents

Introduction to Relative Navigation Concepts for Visiting Vehicles Officers

- Guidance, Navigation, and Control by Flight Phase
- Relative Sensors, Relative GPS, and Ground Tracking
- Integrated Guidance, Navigation, and Control for Orbital Flight
- The Navigation Process
- Kalman Filtering Overview
- State Vector Propagation and Environment Models
- The Covariance Matrix
- Measurement Partials
- Measurement Residuals
- Ratios and Residual Edit Tests
- State Vector Updates, Covariance Updates, and Kalman Gain
- Filter Convergence and Operation
- Sequential Kalman Filtering Versus Batch Filtering
- Summary
- Appendix A – Kalman Filter Derivation
- Appendix B – Measurement Partial Derivations
- Appendix C – Relative Frames

Introduction to Rendezvous Burn Targeting and Guidance Concepts for Visiting Vehicles Officers

- Guidance, Navigation, and Control Overview
- Launch Windows
- Phantom Plane Targeting
- Phasing, Height, Coelliptic, and Planar Control Burns
- The Rendezvous Profile and Constraints
- Targeting a Profile of Burns
- Lambert Targeting
- Clohessy-Wiltshire Targeting
- Closed Versus Open Loop Guidance
- Correlated Velocity Guidance
- Powered Explicit Guidance (PEG)
- Inertial External ΔV Guidance
- LVLH External ΔV Guidance
- Cross Product Steering
- Guidance Functions Before and During a Burn
- Summary
- Appendix A – Time-to-Go and Mass Computation
- Appendix B – Finite Burn Computation
- Appendix C – Basic PEG Thrust Integrals
- Appendix D – PEG Gravity Modeling
- Appendix E – Derivation of the Clohessy-Wiltshire Equations
- Appendix F – Relative Frames
- Appendix G – Burn Nomenclature

Background on These Books

In early 2010 the author was approached by MOD Flight Dynamics Division Visiting Vehicles Officers (VVO) about preparing a version of the shuttle rendezvous training document that was tailored to the needs and interests of the VVOs. Work began on two volumes, *Introduction to Relative Navigation Concepts for Visiting Vehicles Officers* and *Introduction to Rendezvous Burn Targeting and Guidance Concepts for Visiting Vehicles Officers* in the spring of 2010.

Since much technical information concerning Visiting Vehicles (Soyuz, Progress, Automated Transfer Vehicle (ATV), H-II Transfer Vehicle (HTV), Dragon, Cygnus, etc.) was proprietary the author suggested that the new books discuss general concepts, as opposed to vehicle specific details. This would ensure that the books could be made available to personnel working multiple NASA programs. This approach was taken.

Much material from the fourth edition of *Introduction To Space Shuttle Rendezvous Guidance, Navigation, and Control* was used, but most shuttle specific details were removed. Most pages copied from the shuttle book were extensively re-written and additional research was performed. Some Gemini, Apollo, Skylab, and Space Shuttle details were included to illustrate rendezvous and proximity operations concepts.

Compilation Volumes of Rendezvous and Proximity Operations Papers

- Goodman, John L., (editor), *Space Shuttle Rendezvous and Proximity Operations Overview and Experience Papers, Volume 1 of 2 (1970-1985)*, JSC-35050, Flight Dynamics Division, Mission Operations Directorate, NASA Johnson Space Center, July 2011. 633 pages.

- Goodman, John L., (editor), *Space Shuttle Rendezvous and Proximity Operations Overview and Experience Papers, Volume 2 of 2 (1986-2009)*, JSC-35050, Flight Dynamics Division, Mission Operations Directorate, NASA Johnson Space Center, July 2011. 969 pages.

- Goodman, John L., (editor), *Space Shuttle Rendezvous Profile Design Papers, Volume 1 of 2 (1969-1983)*, JSC-35051, Flight Dynamics Division, Mission Operations Directorate, NASA Johnson Space Center, July 2011. 744 pages.

- Goodman, John L., (editor), *Space Shuttle Rendezvous Profile Design Papers, Volume 2 of 2 (1984-2007)*, JSC-35051, Flight Dynamics Division, Mission Operations Directorate, NASA Johnson Space Center, July 2011. 756 pages.

- Goodman, John L., (editor), *Space Shuttle Proximity Operations Design Papers, Volume 1 of 2 (1970-1979)*, JSC-35052, Flight Dynamics Division, Mission Operations Directorate, NASA Johnson Space Center, July 2011. 1145 pages.

- Goodman, John L., (editor), *Space Shuttle Proximity Operations Design Papers, Volume 2 of 2 (1980-2009)*, JSC-35052, Flight Dynamics Division, Mission Operations Directorate, NASA Johnson Space Center, July 2011. 810 pages.

- Goodman, John L., (editor), *Space Shuttle Rendezvous Maneuver Targeting Papers*, JSC-35053, Flight Dynamics Division, Mission Operations Directorate, NASA Johnson Space Center, July 2011. 1178 pages.

- Goodman, John L., (editor), *Space Shuttle Relative Navigation Papers, Volume 1 of 2 (1969-1979)*, JSC-35054, Flight Dynamics Division, Mission Operations Directorate, NASA Johnson Space Center, July 2011. 1244 pages.

- Goodman, John L., (editor), *Space Shuttle Relative Navigation Papers, Volume 2 of 2 (1980-2010)*, JSC-35054, Flight Dynamics Division, Mission Operations Directorate, NASA Johnson Space Center, July 2011. 1056 pages.

- Goodman, John L., (editor), *Gemini Rendezvous Papers*, JSC-35055, Flight Dynamics Division, Mission Operations Directorate, NASA Johnson Space Center, July 2011. 1558 pages.

Approved for public release via STI DAA 24818. See statement on title page.

- Goodman, John L., (editor), *Apollo Rendezvous Papers, Volume 1 – Lunar Missions*, JSC-35056, Flight Dynamics Division, Mission Operations Directorate, NASA Johnson Space Center, July 2011. 1816 pages.

- Goodman, John L., (editor), *Apollo Rendezvous Papers, Volume 2 – Skylab and Apollo/Soyuz*, JSC-35056, Flight Dynamics Division, Mission Operations Directorate, NASA Johnson Space Center, July 2011. 936 pages.

Each compilation volume contains a chapter that provides an overview of the documents, memos, and presentations in the volume. This is followed by chapters listing additional resources on rendezvous (JSC documents, NASA contractor reports, and open literature papers).

Table of Contents for Each Volume

- Series Preface
- Introduction
- Overview of Documents in This Volume
- Sections with references for:
    - Rendezvous Training Documents, Crew Procedures, and Console Handbooks
    - Space Shuttle Flight Software Requirements
    - Additional Resources on Rendezvous
- Documents by Year

Background on the Compilations

During the research and writing of the *Space Shuttle Rendezvous and Proximity Operations Experience Report* from June 2002 to March 2003 a significant amount of primary source material was collected and examined.[1] This material, plus other rendezvous documentation collected by the editor over a period of 20 years, took up five filing cabinets and two bookshelves Included in this material were documents that provided insight into shuttle rendezvous and proximity operations profile evolution, navigation hardware development, burn targeting software architecture, technical challenges, flight experiences, and lessons learned.

In June of 2008, with the end of the Space Shuttle Program approaching, the editor conceived the idea of preserving some of the more insightful documents in a manner that would make them easily available to future engineers and historians. While some primary source material on NASA human flight rendezvous and proximity operations was preserved in existing collections that were easily accessible, many key documents collected by the editor were not. The documentation was sorted and key memos and presentations were assembled into 12 rendezvous and proximity operations compilation volumes

---

[1] Goodman, John L., and Kelli S. Wiuff, *Space Shuttle Rendezvous and Proximity Operations Experience Report*, JSC-49626, Flight Design and Dynamics Division, Mission Operations Directorate, NASA Johnson Space Center, February 2003.

Approved for public release via STI DAA 24818. See statement on title page.

History of Space Shuttle Rendezvous

- Goodman, John L., *History of Space Shuttle Rendezvous, Revision 3*, JSC-63400, Flight Dynamics Division, Mission Operations Directorate, NASA Johnson Space Center, October 2011. 316 pages

This is a revised edition of the *History of Space Shuttle Rendezvous*, updated through the end of the Space Shuttle Program. The original edition, published in October of 2006, was 87 pages long and grew out of a much shorter article that appeared in the *AIAA Journal of Spacecraft and Rockets* at the same time.[1]

The book provides a high level overview of shuttle rendezvous missions, evolution of rendezvous and proximity operations profiles and development of relative navigation and burn targeting sub-systems. The revised edition also includes updated and lengthened chapters covering rendezvous activities of the Gemini, Apollo, Skylab, and Apollo/Soyuz missions. Three new special focus chapters were added to provide additional insight on shuttle rendezvous missions. These chapters include the STS-39 deploy/retrieve mission, missions to the Hubble Space Telescope (HST), and the STS-130 mission to the International Space Station.

A fourth chapter answers the question, "Why was shuttle rendezvous and docking not fully automated?" The chapter covers the levels of automation versus manual control on the Mercury, Gemini, Apollo, and Space Shuttle vehicles. Included is a discussion of why the final approach and landing of the Space Shuttle was manually flown. Appendices list NASA and contractor Mission Control rendezvous team members that supported missions from June 1983 through the end of the Shuttle Program in July of 2011.

Table of Contents

- Preface
- Introduction
- Early Studies
- Mercury
- Gemini
- Apollo
- Skylab
- Apollo/Soyuz Test Project
- Space Shuttle – A New Direction in Mission Activities
- Early Shuttle Rendezvous Studies

---

[1] Goodman, John L., "History of Space Shuttle Rendezvous and Proximity Operations," *AIAA Journal of Spacecraft and Rockets*, Vol. 43, No. 5, Sept.-Oct. 2006, pp. 944-959.

- Shuttle Design Reference Missions
- Plume Impingement
- On-Board Systems
- Skylab Reboost
- Coelliptic versus Stable Orbit Rendezvous
- First Proximity Operations and Rendezvous Flights
- Challenges of Subsequent Rendezvous and Proximity Operations Flights
- Rendezvous or Proximity Operations Technique Demonstration Missions
- Satellite Servicing Missions
- Deployment and Retrieval of Scientific Payloads
- Retrieval and Return to Earth of a Satellite
- Mir and the International Space Station
- The STORRM DTO
- STS-39, the Most Complex Deploy/Retrieve Mission
- A Closer Look at the Hubble Servicing Missions
- STS-130 Mission to the ISS
- Why Was Shuttle Rendezvous And Docking Not Fully Automated?
- Appendices listing Mission Control rendezvous team members.

Guidance and Targeting Reports

Four of the five reports in this chapter concern the algorithms use to target and guide rendezvous burns while the Space Shuttle was acquiring and processing relative navigation measurements during rendezvous (range to the target spacecraft typically less than 40 nautical miles). The fifth report (two volumes) concerns the powered flight guidance algorithm used for ascent, orbit insertion, and deorbit. Four of the reports are compilation volumes that preserve primary source documents in a manner to make them understandable to future generations of engineers.

Lambert Targeting Compilation

- Goodman, John L., (editor), *Space Shuttle Lambert Targeting*, JSC-49708, Flight Design and Dynamics Division, NASA JSC, May 2003. 336 pages.

This is a compilation of reports and presentations on the theory behind the rendezvous burn Lambert targeting algorithm used in the shuttle flight software. Some of the material was collected and used during investigation of the STS-49 (May 1992) Lambert targeting anomaly.[1]

Lambert Cyclic Guidance Compilation

- Goodman, John L., (editor), *Space Shuttle Lambert Cyclic Guidance*, JSC-49709, Flight Design and Dynamics Division, NASA JSC, May 2003. 556 pages.

This report is a collection of documents concerning closed loop Lambert cyclic guidance, used by the Space Shuttle for rendezvous burns targeted with the on-board Lambert targeting algorithm. The algorithm was a form of correlated velocity guidance. It used a burn position biasing equation developed by Tim Brand of Draper Laboratory and discussed by Dr. Richard Battin at the end of the introduction to *An Introduction to the Mathematics and Methods of Astrodynamics*.[2] The report contains memos detailing the STS-69 Lambert cyclic guidance performance investigation.[1]

STS-49 Lambert Targeting Compilation

- Goodman, John L., (editor), *STS-49 Lambert Targeting Anomaly and Aftermath*, JSC-49710, Flight Design and Dynamics Division, NASA JSC, May 2003. 640 pages.

This report contains extensive documentation of the STS-49 (May 1992) Lambert targeting anomaly and analysis conducted in support of the investigation.[1] Also included are memos and presentations concerning a Lambert targeting software problem discovered five days before the flight of STS-51 in September of 1993.[1]

---

[1] Goodman, John L., *Lessons Learned From Seven Space Shuttle Missions*, NASA Contractor Report NASA/CR-2007-213697, NASA Johnson Space Center, January 2007.

[2] Battin, Richard H., *An Introduction to the Mathematics and Methods of Astrodynamics*, Revised Edition, American Institute of Aeronautics and Astronautics, Reston, VA, 1999.

Lambert Cyclic Guidance Upgrade

- Meissen, Thomas J., *Space Shuttle Lambert Guidance Improvement, SCR 92843/93009 OI-32*, JSC-49830, NASA/JSC Flight Design and Dynamics Division, October 28, 2006. 84 pages.

In the aftermath of the STS-69 Lambert cyclic guidance performance anomaly on STS-69 (September 1995) John Goodman developed and tested an upgraded version of Lambert guidance.[1,2] This upgrade was later approved for incorporation into the flight software. The report by Thomas Meissen details the engineering work performed to transform the upgrade into a shuttle flight software change.[3] The first flight of the upgraded Lambert guidance algorithm (flight software version OI-32) was the STS-120 (October-November 2007) mission of *Discovery* to the International Space Station.

---

[1] Goodman, John L., *Lessons Learned From Seven Space Shuttle Missions*, NASA Contractor Report NASA/CR-2007-213697, NASA Johnson Space Center, January 2007.
[2] Goodman, John L., (editor), *Space Shuttle Lambert Cyclic Guidance*, JSC-49709, Flight Design and Dynamics Division, NASA JSC, May 2003.
[3] On September 13, 2005 the report was provided to a Boeing engineer who was responsible for testing the guidance upgrade in the Shuttle Avionics Integration Laboratory (SAIL) in JSC/Building 16. The engineer had no previous experience with Lambert guidance. The response upon examining the report was, "Oh, boy, I think I hit the mother lode!! I need to buy a lottery ticket today."

Approved for public release via STI DAA 24818. See statement on title page.

Compilation of Powered Explicit Guidance Papers

- Goodman, John L., (editor), *Compilation of Powered Explicit Guidance Papers (1971-1980)*, Volume I, JSC-64694, NASA JSC Flight Dynamics Division, February 2010. 1032 pages.

- Goodman, John L., (editor), *Compilation of Powered Explicit Guidance Papers (1981-2010)*, Volume II, JSC-64694, NASA JSC Flight Dynamics Division, February 2010. 992 pages.

These volumes preserve for future generations of engineers 91 documents created over a period of 40 years concerning Space Shuttle Powered Explicit Guidance (PEG). Although the initial theoretical development of the vector Linear Tangent Guidance (LTG) predictor corrector algorithm that forms the basis of PEG occurred in about 1970, PEG is still a state-of-the-art guidance algorithm.[1] PEG performed flawlessly over the life of the Shuttle Program for nominal ascent, orbit insertion, deorbit, and aborts. PEG was also selected for orbit insertion and deorbit guidance of the Orion vehicle, and was chosen by the Marshall Space Flight Center for Ares I guidance. PEG was used for lunar landing and ascent guidance in the NASA Autonomous Landing and Hazard Avoidance Technology (ALHAT) project in the Constellation Program and other lunar landing studies.

These volumes cover three phases of shuttle guidance history. The first is 1971 and 1972 (Volume I), when studies were performed on potential shuttle guidance algorithms. The second phase (Volume I) covers 1973 through 1980 when the basic PEG theory was further developed and adapted into a workable Shuttle guidance algorithm covering powered ascent, aborts, orbit insertion, and deorbit. The third phase (Volume II) covers 1981 through 2010 and consists mostly of educational memos and presentations on PEG along with some material on PEG performance improvements.

Both volumes include the memories and commentary of three engineers who participated in PEG development and improvement in the 1970s and 1980s. Footnotes were included by the editor to provide additional background information. Additional sections list other resources for shuttle powered flight guidance, such as training manuals, software requirements, and other JSC documents.

Table of Contents

- Introduction
- Suggested Reading List
- References for Overview Papers
- Descriptions of Papers in the Volumes

---

[1] Jaggers, Roland F., "An Explicit Solution to the Exoatmospheric Powered Flight Guidance and Trajectory Optimization Problem for Rocket Propelled Vehicles," *AIAA Guidance and Control Conference*, Hollywood, FL, August 8-10, 1977.

- Comments by PEG Development Veterans
- Shuttle Ascent Training Documents
- First Stage Ascent Guidance
- Flight Software Requirements
- Other Relevant JSC Documents
- 91 PEG documents in chronological order

Apollo and Orion Reports

The three conference papers in this chapter were written in support of the Orion Program. Two are specifically about Orion, while the Apollo 13 paper was written to provide Orion engineers with lessons learned from that mission. An overview of Apollo entry guidance was written by an Apollo veteran for Orion personnel.

Apollo 13 Guidance, Navigation, and Control Challenges

- Goodman, John L., "Apollo 13 Guidance, Navigation, and Control Challenges," *AIAA Space 2009 Conference & Exposition*, September 14-17, 2009, Pasadena, CA. 42 pages.

This paper was written to specifically address guidance, navigation, and control (GNC) aspects of the Apollo 13 mission for personnel working on Orion contingency return to Earth techniques and GNC design. Books written for the general public and NASA reports on the mission focus on the life support, power, thermal control, and human interest aspects of Apollo 13. The paper was written based on research using the crew debrief transcript, the air-to-ground communications transcript, and post flight reports written by Mission Control and Mission Planning and Analysis Division (MPAD) personnel immediately after the flight.

Topics include the nominal mission and lunar trajectory plan, nominal crew activities, GNC performance before the oxygen tank incident, flight control challenges immediately after the incident until Lunar Module (LM) activation, LM Inertial Measurement Unit (IMU) alignments using the Sun instead of stars, re-establishing the return to Earth trajectory, trajectory correction burns, use of the Earth as a pointing cue to achieve burn attitudes, ground based orbit determination challenges, maneuvers to burn attitudes, LM and Service Module separation from the Command Module, separation trajectories, re-entry, and landing. Lessons learned from Apollo 13 as related to GNC and ground support personnel are detailed. The GNC architectures and hardware components of the Lunar and Command/Service Modules are discussed in an appendix.

Introduction to Apollo Entry Guidance and Flight Performance

Burton, John K., *Introduction to Apollo Entry Guidance and Flight Performance*, JSC-35020, NASA JSC Flight Design and Dynamics Division, July 2009. 62 pages.

This document was written by Apollo and Space Shuttle entry guidance veteran John Burton (formerly of United Space Alliance) to introduce Constellation personnel to Apollo entry guidance. It includes discussion of theoretical entry guidance developments before Apollo, Primary Guidance, Navigation, and Control System (PGNCS) entry guidance, the backup Entry Monitoring System (EMS), and the entry guidance performance footprint. Excerpts from Apollo mission reports concerning re-entry performance are also included.

Table of Contents

- Introduction
- Early Development of Entry Guidance
- Theory and Technique
- Mercury Entry
- Gemini Entry
- Apollo Entry
- The Apollo Spacecraft
- Overview of Apollo Entry Guidance
- Apollo Entry Corridor
- Skip Entries vs. Direct Entries
- Entry Guidance Performance Footprint
- Details of the Apollo Guidance Algorithm
- Entry Monitoring System (EMS)
- Backup Trajectory Control Modes
- 1972 Apollo Entry Guidance Experience Report
- Apollo Entry Guidance vs Orion PredGuid
- Summary
- Appendix A: Un-manned Flight Testing of the Apollo Spacecraft
- Appendix B: Manned Apollo Flights
- Appendix C: Atmospheric Entry in Foreign Programs
- Acronyms
- References

Challenges of Orion Rendezvous Development

- Goodman, John L., J. P. Brazzel, and D. A. Chart, Ph.D., "Challenges of Orion Rendezvous Development," *AIAA Guidance, Navigation, and Control Conference*, AIAA, Reston, VA, 2007. 30 pages.

The paper "Challenges of Orion Rendezvous Development" details differences between Orion and previous NASA human spacecraft (Gemini, Apollo, and the Space Shuttle) in the areas of rendezvous, proximity operations, and docking. Differences discussed include mission requirements and Orion sub-systems. Orion had requirements for a higher level of automation and autonomy during rendezvous and docking than previous NASA human spacecraft. Automation and autonomy are discussed in detail, as well as the importance of balancing levels of authority (crew and ground), automation, and autonomy so that mission success and crew safety is ensured.

Topics Covered

- Introduction
- Programmatic Challenges
- Rendezvous from Gemini to Orion
- Orion Rendezvous
- Automation and Autonomy for Orion
- Conclusion
- Appendix – Historical Summary of Automation in NASA Human Flight Programs

Orion On-Board Navigation Architecture and Operations Concepts

- Goodman, John L., H. R. Mamich, and D. W. Saley, "Orion On-Board Navigation Architecture and Operations Concepts," *Guidance and Control 2008, Advances in the Astronautical Sciences*, Volume 131, Univelt, San Diego, CA, 2008, pp. 425-444. 20 pages.

This paper describes the Orion navigation architecture, sensors, and operations concepts as of February of 2008. The Orion navigation team was tasked to design a navigation architecture incorporating advanced technology and new operations concepts, while facing challenges of low technological readiness levels, crew safety, and limitations in vehicle weight, available power, budget, and schedule. While Orion was to fly the same flight phases as Apollo, there were important navigation differences between the spacecraft. Optical tracking of lunar landmarks and stars was to be performed automatically, rather than using Apollo era human-in-the-loop sextant tracking. On-board GPS orbit determination could be used in low Earth orbit to supplement ground radar tracking. GPS technology would be used to fly long-range lunar return skip entry trajectories that were not flown during crewed Apollo missions. Orion navigation was more automated than navigation systems on previous NASA human flight vehicles and would require less crew interaction.

The paper discusses the evolution of the Orion navigation system design and changes that occurred during the design phase. Changes to the navigation system stemming from the 2007 vehicle weight reduction exercise are detailed. Back-up and emergency navigation capabilities are discussed, along with fault detection, isolation, and recovery. The navigation operations concepts during the pre-launch, nominal ascent, ascent abort, Earth orbit, low Earth orbit rendezvous, transit to and from the Moon, low lunar orbit, lunar orbit rendezvous, nominal atmospheric entry, and entry following an ascent abort are detailed.

Compilations of Draper Laboratory Papers

- Goodman, John L. (editor), *Charles Stark Draper Laboratory Apollo Papers*, JSC-35085, Flight Dynamics Division, Mission Operations Directorate, NASA Johnson Space Center, June 2011. 1712 pages.

- Goodman, John L. (editor), *Charles Stark Draper Laboratory Space Shuttle Papers*, JSC-35083, Flight Dynamics Division, Mission Operations Directorate, NASA Johnson Space Center, June 2011. 1739 pages.

These volumes contain memos, formal reports, conference papers, and presentations covering the years 1959-2010 by personnel from the Charles Stark Draper Laboratory (before 1970 called the MIT Instrumentation Laboratory). The first document covers the Apollo and Skylab Programs, while the second covers the Space Shuttle Program. Draper Lab made significant guidance, navigation, and control contributions to both programs. The reports preserved in these volumes are an important source of information on guidance, navigation, and control algorithms and techniques.

Summary

The motivation to write the reports and the knowledge capture approach used was based on experiences with Shuttle Program corporate knowledge loss and difficulties encountered during the Orion Program with technical history research into Apollo and the Space Shuttle. Some of the reports were written for internal knowledge capture and training, other were written to transmit lessons learned and experiences to external audiences, and other reports and document compilations were created during the knowledge capture efforts near the end of the Space Shuttle Program. Subject matter experts with proficiency for research and written communication were used as report authors to ensure that quality reports were created in a timely and low cost manner. Complete paragraphs and sentences rather than bullet points and spreadsheets were used to ensure effective knowledge transfer to current and future engineers. Effective knowledge capture to prevent corporate knowledge loss and educate future engineers is necessary to reduce technical, cost, schedule, mission success, and flight safety risk during both vehicle development and mission execution.

This page intentionally left blank.

Distribution

NASA Johnson Space Center
AD1/ Lulla, Kamlesh
AK / Andrews, Cheryl
AK / Engle, Jeanie
AK / Fontenot, Brent
DA111 / Beck, Kelly
DA111 / Hill, Paul
DA111 / Koerner, Stephen
DA111 / Schaefer, Stan
DA33 / Demaret, Martin
DA335 / MOD Library
DA8 / Dye, Paul
DA8 / Edelen, Chris
DD111 / Sims, John
DM / Banfield, Rob
DM / Gavin, Richard
DM / Fields, John
DM / Library
DM3 / Reichert, Christine
DM32 / González, Edward
DM34 / Dannemiller, Dave
DM34 / Spehar, Jeannette
DM34 / Walker, Steve
DM42 / Holt, Greg
DM46 / Powers, William
DS111 / Azbell, Jim
DS21 / Dunham, Scott
EA111 / Chapline, Gail
EA421 / Wagner, Howard
EG1 / Labbe, Steven
EG1 / Straube, Tim
EG4 / Fitzgerald, Steve
EG5 / Cerimele, Chris
EG5 / Hoffman, David
EG5 / Whitley, Ryan
EG6 / Brazzel, Jack
EG6 / Crain, Tim
EG6 / D'Souza, Chris
EG6 / Mitchell, Jennifer
EG6 / Mrozinski, Richard
EG6 / Spehar, Pete
EV151 / Murray, Scott
IM111 / Scroggins, Mark
IP111 / McGuyer, Michael
IS / George, Christa
IS20 / Ross-Nazzal, Jennifer
IS20 / Wright, Rebecca
IS23 / STI Center (Caballero) (4)
MG111 / Wooten, Peggy
MS111 / Murphy, Terri
MS111 / Riojas, Dalia
NA131 / Erminger, Mark
OE111 / Lutomski, Mike
SA4 / Lane, Helen
ZD / Pearson, Don
ZF3 / Goforth, Monty
ZV / Mulligan, Dan

NASA Ames Research Center
Moffett Field, CA 94035
MS 202-3 / Library
NASA Glenn Research Center
21000 Brookpark Road
Cleveland, OH 44135
105-3 / Burke, Laura
105-3 / Sjauw, Waldy
142-3 / Technical Library
NA00 / Marinaro, John
OAI-NSC / Lilley, Steve

NASA Goddard Space Flight Center
Greenbelt, MD 20771
100 / Fillip, Barbara
272 / Banholzer, Gordon
564 / Katz, Richard
590 / Dennehy, Cornelius
595 / Barbee, Brent
595 / Berry, Kevin
595 / Carpenter, Dr. Russell
595 / Gaebler, John
595 / Getzandanner, Kenny
595 / Moreau, Michael
595 / Naasz, Bo
1000 / Rogers, Edward

NASA Headquarters
Washington, DC 20546
BA000 / McAllister, Phil
BJ000 / Lengyel, David
CG000 / Vrotsos, Pete
KA0 / Bauer, Frank
NH / Barry, William
NH / Garber, Stephen

NASA Kennedy Space Center
Kennedy Space Center, Florida 32899
NE-A2 / Menendez, Alfred
Library-D

NASA Langley Research Center
Hampton, VA 23681
105 / Pahlavani, Patricia
185 / Document (Miller, Sue)

NASA Marshall Space Flight Center
Huntsville, AL 35812
ES32 / Bryan, Thomas
ES32 / Howard, Richard
EV42 / Heaton, Andrew
EV94 / Kulpa, Vyga

United Space Alliance-Houston
USH-400A / Hartwig, Scott
USH-420A / Judd, Bryan
USH-482L / Brockway, Dan
USH-483L / Hill, Richard
USH-484T / Frank, Jorge
USH-800A / Osterlund, Jeffrey
USH-806A / Goodman, John
USH-806A / Thomas, Mark

Jet Propulsion Laboratory
4800 Oak Grove Dr.
Pasadena, CA 91109
111-113 / Library
238-600 / Bertiger, Willy
301-250D / Equils, Douglas

Department of the Army
US Army Research, Development and Engineering Command
Redstone Scientific Information Center
Building 4484
Redstone Arsenal, AL 35898
RDMR-CSI/Knott